Ship of Thieves

Douglas Skelton has published numerous non-fiction books and crime thrillers. He has been a bank clerk, tax officer, shelf stacker, meat porter, taxi driver (for two days), wine waiter (for two hours), reporter, investigator and local newspaper editor. He has been longlisted for the McIlvanney Prize five times, most recently in 2024. Douglas contributes to true crime shows on TV and radio and is a regular on the crime writing festival circuit.

Also by Douglas Skelton

A Company of Rogues

An Honourable Thief
A Thief's Justice
A Grave for a Thief
A Thief's Blood
Ship of Thieves

Other novels by Douglas Skelton

Blood City
Crow Bait
Devil's Knock
Open Wounds
The Dead Don't Boogie
Tag – You're Dead
The Janus Run
Thunder Bay
The Blood is Still
A Rattle of Bones
Where Demons Hide
Children of the Mist
The Hollow Mountain

DOUGLAS SKELTON

Ship of Thieves

CANELO

First published in the United Kingdom in 2025 by

Canelo, an imprint of
Canelo Digital Publishing Limited,
20 Vauxhall Bridge Road,
London SW1V 2SA
United Kingdom

A Penguin Random House Company
The authorised representative in the EEA is Dorling Kindersley Verlag GmbH. Arnulfstr. 124, 80636 Munich, Germany

Copyright © Douglas Skelton 2025

The moral right of Douglas Skelton to be identified as the creator of this work has been asserted in accordance with the Copyright, Designs and Patents Act, 1988.

All rights reserved. No part of this publication may be reproduced or transmitted in any form or by any means, electronic or mechanical, including photocopy, recording, or any information storage and retrieval system, without permission in writing from the publisher.

No part of this book may be used or reproduced in any manner for the purpose of training artificial intelligence technologies or systems. In accordance with Article 4(3) of the DSM Directive 2019/790, Canelo expressly reserves this work from the text and data mining exception.

A CIP catalogue record for this book is available from the British Library.

Print ISBN 978 1 80436 739 1
Ebook ISBN 978 1 80436 744 5

This book is a work of fiction. Names, characters, businesses, organizations, places and events are either the product of the author's imagination or are used fictitiously. Any resemblance to actual persons, living or dead, events or locales is entirely coincidental.

Cover design by Henry Steadman

Cover images © Shutterstock

Printed and bound in Great Britain by Clays Ltd, Elcograf S.p.A.

Look for more great books at
www.canelo.co | www.dk.com

In memory of my friend Denzil Meyrick

Greatly missed

I'm sure he's reading this somewhere and pointing out that I didn't call him a genius

Prologue

Edinburgh, January 1718

1

The last thing Gideon Flynt expected on a cold, damp and windy High Street was to come face to face with a dead man.

He was walking back from the Grassmarket to his tavern, his shoulders hunched against the chill, his hands thrust into the pockets of the greatcoat he had worn since his early days at sea. There were times when he felt the call of strong liquor, when the melancholy came upon him and he had a need to seek his own company but didn't think it fitting that he imbibe in his own establishment. When such a desire came upon him, the landlord of the White Horse, in a display of courtesy to a fellow tavern owner, made a private room available. It wasn't the drink that helped, it was the solace of solitude, a time for his mind to settle, for his spirits to lighten.

He didn't know what caused those infrequent periods of introspection but when they did come upon him his wife, Mercy, understood and left him to his own devices. An hour, perhaps two, in that little room, alone apart from the innkeeper who served him personally, and the gloomy mood would lift and he could return to her.

The bitter chill besieged him as he stepped from the warmth of the tavern, the air filled with a moisture that wasn't quite rain but formed into a damp mist so subtle that it could penetrate cloth and skin to seep into the soul. It haloed the lanterns outside the doorways and softened the edges of the buildings clustered around the Grassmarket. Nary a soul ventured abroad in such dismal weather and truly this was a night to have stayed by the fireplace in his tavern, listening to the conversations of the men and women around him: talk of politics, of personal domestic issues, of events of the day and, yes, in the darker corners of the room, where the shadows cast by the tallow candles were deeper, even a little sedition. As he trudged up the West Bow that linked the low town to the high, he eyed the windows over the cobblers' workshop, part of him hoping to see a candle aflame, but they were dark. His stepdaughter, Cassie, would be abed by now, he reasoned, but he would have dearly loved to have called upon her, to both warm his bones before her fire and to talk out the dregs of his low mood instead of trying to drown it, for sometimes it could hold its breath even longer than a pearl diver in the Japans. Cassie had a way of seeing the nub of an issue and tackling it head-on. She would recognise the right and wrong of it and would point it out to him in no uncertain terms. There was no issue, however; it was Gideon Flynt's own regrets and failings – even his guilt – coming to the surface.

He lumbered on, his bones and muscles reminding him that he was no longer the cocky whelp who had left the old town to find adventure at sea. He was older now, slower of body if not of reason, and the rise of this crooked street seemed so much more difficult. As they said in this town, old age does not come alone. It was as he reached the angle of the Bow, where the street turned right before twisting left again, left to head uphill to the Lawnmarket, that he spied two men alighting from a carriage. They were wise men, well prepared for the cut and slice of an Edinburgh winter. Their coats were sturdy, their hats low against the tiny droplets of water that hung in the air. They were the only souls he saw in his walk. Even the area around the Weighhouse at the top of the Bow was deserted, and it was here that the wind sweeping up from the Forth caught him and cut at him like a cutlass. When the climate was kinder there might have been a few loungers sharing a bottle or a caddy or two looking for strangers to convey to lodgings or for whom to convey a message, but this night there were none. Not even a Town Guard walked the precinct from the High Street to the castle, they being a not-so-hardy race of men who preferred the warmth of their guardhouse. The Town Cross was another rallying point for caddies but also chairmen, though he would lay good silver down that even those hardy souls had deserted it on this night, there being far from a glut of commerce. Gideon gave hardly a second look to the dark, brooding hulk that was the Tolbooth, the town jail, positioned in the street like a warning to any malefactors against proceeding further. He was well-acquainted with that tall building's stern exterior and it held no fascination for him. Instead, his attention was taken up by the carriage lingering by the opening to the passageway leading to the courtyard in which his tavern was situated.

And then he heard his wife's voice, crying out in rage, and the distinctive sound of a blow being struck.

A man cursed, his voice a bellow of pain, and it was followed by another, heavier blow, and Mercy moaned. Gideon broke into a run just as a figure stepped into his path from the rear of the carriage.

'Hello, Gideon.'

The voice was an echo from the past, but Gideon didn't at first put a name to it. He strained in the half-light of the lanterns to make out the man's features but they were obscured by a scarf and the shadow cast by the brim of his hat. Two men dragged Mercy's inert frame from the close and manhandled her into the carriage. Gideon stepped towards them but the man raised a pistol and held it steady in his direction.

'Don't,' was all he said.

'That's my wife,' Gideon said.

'That's my property,' said the man.

'Your...' Gideon was confused but then understanding dawned, confirmed by the man unwrapping the scarf from his face with his free hand.

'Toby Hawke,' Gideon said.

'The very same,' the man said, with a smile that was more vicious than welcoming.

'But I...'

'You thought me dead, your blade between these ribs,' Hawke said, slapping his chest. 'But as you can see, you didn't do the job proper. The devil didn't take me but I'm here to take what is mine, rightful and legal.' He flicked the pistol barrel quickly towards the coach to his right, into which Mercy had been delivered. 'And we'll have her cub soon, too.' He smiled again at observing Gideon's reaction. 'Oh yes, we know where she is, and I have men there even now. She's a beauty, and I'm sure they'll have some fun with her, too.'

Gideon's mind jumped to the two men emerging from the coach near to Cassie's home.

'I'll kill you,' Gideon said, his body straining to dart forward.

'You tried that once before and failed. Perhaps you will have better luck this time.' Hawke waved the weapon again. 'Perhaps not, for this time you cannot strike at me unawares.'

The gap between them was too great and Gideon was no longer young. But then, neither was Hawke. He had aged, and badly. He had once been handsome, even Gideon had recognised that, but his cheeks had hollowed and his eyes had sunk into their sockets, giving him a death's head appearance. He wondered if the knife blow he had struck all those years ago – so many years ago – had caused that deterioration, or was it his own vile nature that had eaten at him? One of his men climbed into the seat and took the reins of the carriage, looking backwards over the roof to watch. Gideon had to steal the opportunity to think, to plan a move.

'Why, after all this time, Hawke?'

'Time doesn't remove my property rights, Gideon. I will have what is mine returned. And repay my debt to you, too. It took me a time to heal and you had laid a fine false trail. The lies you told even your shipmates about your life were so complete that I had no idea you came from this here town.'

'Then how did you find me?'

'Blind luck, in truth. I was here on business and saw you in the street. Simple as that. I recognised you, of course, but you did not me. I've changed a great deal these twenty-five years, while you remain much the same. The hair is whiter, the gait slower, the belly a little rounder, but I knew you straight off.'

Gideon's mind raced, trying to formulate a plan, anything that would wrest Mercy from this man's hands. He had done it once before; he could do so again. He just needed time. 'What sort of business does a man like you have in Edinburgh?'

'Business that means nothing to you.'

The man in the box seat grew impatient. 'Father, we must be away...'

So that was Daniel, Hawke's lad. He had only been a bairn when Gideon had seen him last but now that he gave him a brief study, he saw the same handsome features that the father had once borne, the same cruel eyes.

'Aye,' said Hawke. 'It has been a delight to see you again, Gideon, but I must cut our reunion short. I would leave you with a dagger in your chest, as you did me, but old though you are, I don't propose to move any closer than this.'

'You mean to shoot me? This isn't Nassau, this is Edinburgh.' Gideon waved his hand around him. 'There are eyes everywhere, eyes that will see and tongues that will wag. You wouldn't get far before you were intercepted.'

Hawke hesitated, his eyes darting around the street and up at windows. Gideon primed his body to throw himself to the side in what he knew in his heart would be a vain attempt to avoid the ball, then heard voices heading their way. Through the fine mist from the direction of the guardhouse came a brace of Town Guards, their long Lochaber axes propped upon their shoulders, the folds of their heavy russet coats flapping against their legs. They paid the carriage and two men little heed, being so deep in their own converse.

Hawke swore, concealed the pistol in the folds of his greatcoat, but kept it pointed towards Gideon. 'I would have vengeance, but perhaps it's better this way, Gideon. What I do to that creature and her offspring will eat at your mind, at your very soul.' He hauled himself into the coach and cried up to his son, 'Fly, Dan, we must be for Leith.'

The younger Hawke cracked a whip and the pair of horses started immediately down the High Street. Gideon waved his arms to the Guards. 'Stop them, they've abducted my wife.'

The guards, both elderly men, Highlanders probably, came to a halt, not knowing what to do as the carriage rattled towards them. Had they been armed with muskets, had they even been of sufficient dexterity to utilise the hook at the end of the axe's long staff, then perhaps they might have been able to bring the carriage to a halt. All they did was dart from its path as it thundered over the cobbled roadway down the High Street. Gideon could not hope to catch it, so he whirled and ran back towards the Bowhead, ignoring the cries of the guards demanding answers, his only thought now being of Cassie and the two men he had seen earlier.

—

Cassie had already taken herself to bed when she heard the noises in the shop below. They had managed to gain access without any sound but no doubt being unfamiliar with the interior, and it being dark, had blundered into something to send an item or two of equipment clattering to the floor. If there was damage, Mr Shaw, the man who rented the shop from her, would be most displeased. He was very particular concerning the condition of his tools.

On her eyes flying open, she first experienced a jolt of fear, giving way to anger that someone might feel it acceptable to break in. The shoemaker business might no longer be hers, but she still felt proprietorial. If they thought her a helpless woman, they would soon be shown otherwise. The cold assaulted her as soon as she climbed from under the quilt, so she pulled a coat from the nail behind the door and hauled it over her thick cotton night-chemise, then fished a pistol from where she had it hidden under the bed. She was a young widow living with only her son, and though he was growing into a strapping lad and Edinburgh was civilised, it could be far from peaceful. She had herself witnessed riots and tulzies in the streets, even the murder of a man in the name of justice on the Grassmarket for which nobody paid any sort of penalty, and so she kept the weapon cleaned and primed at all times. She knew not the intent of whoever was below, but she wasn't about to simply let them have their way. Cocking the hammer, she eased the door open and stepped into the joint parlour and kitchen to listen at the door. She thought she heard a muffled oath, the voice not Scottish, and then the first footfall on the steps leading from below. Two intruders, she estimated, while she only had one shot, so unless one stood obligingly close to the other so that it would allow her pistol ball to pass through both, she would still have one to deal with. No matter; if she was to do anything, it had to be done while they were on the narrow staircase and she was above them.

The door to the small chamber at the rear of the apartment opened to reveal her son still in his night attire, a candle lighting his way. 'I heard something below,' he whispered. His gaze fell on the pistol in her hand but he made no comment. He seemed calm, and for that she was proud, even if it was perhaps an unwelcome indication of his father's blood, but the candle he held did waver as his hand trembled.

'We have intruders, Jonas,' she said, her voice as low as his as she waved him to remain where he was.

She gripped the door handle, allowed a moment for her own hand to stop shaking. She took a deep breath. Steady your nerves, Cassie, she told herself. You have the advantage of higher ground and surprise on your side. You can take one, and the other may flee, knowing the bark of the pistol will alert neighbours. You can win this. *My God*, she realised, *perhaps some of Jonas Flynt has rubbed off on me, too.*

Freeing her breath through her teeth, she wrenched the door open, stepped onto the small landing pistol first, and fired at the dark bulk of the man nearest her. He cried out, jerked against the wall, both hands darting low to his side. Cassie gripped the pistol by its spent barrel, prepared to use it as a club.

'I don't know who you are, but you've picked the wrong woman to molest,' she said, her voice calm, but she knew the depth of her terror.

'She near got me in my nethers,' said the wounded man, his voice a whine.

'And I'll beat out what brains you have if you come any closer,' she promised.

The second man had little appetite to proceed any further, for he was already backing down the stairs. His companion saw him and cursed him for a coward. 'Get the bitch,' he said.

'You get the bitch,' said the other, already back on the shop floor.

'I'm wounded, for God's sake.'

'Then come back down and let us flee, because this jig is up.'

The man below vanished, heading for the door but the wounded one remained, staring at Cassie as if gauging her resolve to club him down. It was in that brief moment of study that her son burst out from behind her, a heavy iron poker he had lifted from the fireplace raised, and charged down the steps between them. Instinctively she reached out to stop him, fearful that he would come to harm, for that she couldn't bear, but he was already past her. With a roar, he swung his makeshift weapon, aiming for the head but catching the man where the neck met the shoulders. The man's coat cushioned the force of the blow but the shock of the offensive caused him to lose his footing and he tumbled backwards to sprawl half on the bottom two steps and half on the ground floor. Jonas advanced further, ready to strike again.

'No, Jonas!' she cried.

She needn't have worried, for the man showed some considerable agility in pulling himself to his feet and limping away in retreat, pursued by her son, still yelling as if he was going into battle.

'Leave him,' she shouted, gripping him by the shoulder to tug him backwards, lest others lay in wait in the street. She peered out and saw the wounded man being assisted into a coach waiting in the wet mist. A whip cracked and the vehicle clattered past them towards the Grassmarket. Cassie caught sight of the wounded man's expression, a mix of pain and rage, as he glared at them.

Jonas watched it go, his face taut. 'Who were they?'

Her heart hammering, her hand shaking with the release of tension, she said, 'I don't know.'

'What did they want?'

'I don't know that either.' She laid a hand on his shoulder. 'You were very brave.'

He smiled. It was a bashful smile, though, as if he suddenly realised what he had done. His fingers trembled as he ran a hand through his hair. 'It was all by instinct.'

She knew where that instinct came from, much as she didn't wish to go there. Jonas knew the truth of his father, was aware that the man he had thought sired him had merely raised him, and he had come to accept it. He was growing into a fine young man, tall, strong, but studious. Unlike Jonas Flynt, his true father, he was studying law, not breaking it. She gently urged him back into the shop, aware that the bitter chill would freeze him ere long.

The pound of footsteps down the West Bow made her stiffen, preparing for another attack, but it was Gideon making a swift study of them as he emerged from the gloom. 'You are both unmolested?'

'We are, but that's more than I can say for one of the intruders,' Cassie said. 'Who were they?'

He glanced pointedly at his grandson and gave her a shake of his head. 'There is no time. I must away to Leith.'

He was already turning away and hurrying back up the hill. Cassie called after him. 'What is in Leith?' He didn't reply. 'Gideon, who were those men?'

But he had already merged with the thickening mist and was gone, only the sound of his steps retreating giving testament that he had ever been there.

Part One

London – two weeks later

2

They had been dancing, he and Belle St Clair. She had once pointed out to him that they had never danced, but this night, they did. Jonas Flynt was clumsy and awkward but he did his best not to trip over his own feet. He could be most adroit when in a fight, but on this small area of rug in the parlour of Mother Grady's house on Covent Garden, while Mr George Frideric Handel led a quartet of musicians in his 'Sarabande', he was out of his depth. He had been comfortable, for want of a better term, with the minuet, but this slower-paced dance revealed him to be a lumbering oaf. He experienced a measure of comfort when Colonel Charters led Mother Grady by the hand and joined them in the cramped space.

Belle laughed when she saw the relief in his expression and he laughed with her. She was happy, and he was glad to see it, glad that he could in some way bring her joy. She leaned closer and whispered, 'I never want to lose you.'

He held her gaze. 'You won't, Belle, not if I can help it.'

She understood that his life did not lend itself to making promises of longevity, but he was sincere in his resolution to live as long and as content a life as he could, given the pressures of the work he did for Colonel Charters. He knew now that he wanted to live that life with Belle. When he thought a man with a knife was about to kill her, he had felt a stabbing in his heart that he hadn't experienced for a long time. Guilt he was used to, stirring of conscience also, but the prospect of losing her was something he couldn't bear.

'I love you, Jonas Flynt,' she said.

Nobody had said that to him before. He had never said that to anyone. Now's the time, Jonas, say it. Three words. Say it.

'Belle, I...'

It was at that moment that Jerome, the house guardian, tapped him on the shoulder. 'Begging thy pardon, Mr Flynt.'

Flynt almost groaned with exasperation at the interruption, but Belle laughed. 'Salvation, Jonas,' she said.

'What is it, Jerome?' Flynt asked, his tone sharper than he meant it to be.

'There's some people asking after thee. I tried to explain that thee was otherwise engaged but they would brook no refusal. I put them in t' small room across t' hall.'

Belle laughed again. 'Go and see who it is, Jonas.'

Sighing and giving her an apologetic look, and shrugging in reply to Charters' unspoken query as he danced with Mother Grady, Flynt followed Jerome to the small reception room. Jerome opened the door and stepped aside to let him enter.

His heart began to hammer and his brain froze as the woman turned to face him, a young man at her side. It had been nearly three years since Flynt had last seen him and he had grown considerably into a tall, if slim, young man. His features had also formed much further and, to Flynt, it was like gazing into a looking glass at himself at the boy's age. The boy watched him carefully.

'Jonas,' said Cassie, as ever wasting no time, 'I need your help.'

That astonished him. Cassie had made it clear that she required no help from him.

'It's Gideon,' she added.

That sharpened his attention. 'What's happened?'

She told him what had occurred in Edinburgh. Her words erupted from her, as if they had been desperate for air. Once she had finished, she took a seat in a plush wingback chair, her son stationing himself behind her. As he listened to Cassie's story, Flynt's eyes were drawn to him time and again. There was an intensity in the gaze that he found discomfiting. There was resentment there. There was mistrust. And Flynt couldn't blame him. Those looks flying across the room proved that the boy knew the truth of his parentage.

Mr Handel's music drifted from the parlour into this small room off the foyer as Cassie talked and Jonas glared. Fresh discomfort afflicted him. Belle was in that parlour, resplendent in the gown she had worn for what she had thought was a night at the theatre, only to be surprised by the function Flynt had secretly arranged. Belle, who had told him that she loved him, a sentiment he almost returned. Belle, his new love, and Cassie, his first love, under the same roof. His emotions were in turmoil. He was a man who had faced death on many occasions, had come close to it a few times, but who had always won through. But he was also a man ever daunted by his own deeper feelings.

So he focused on Cassie's narrative, but his mind swirled under the accusatory glare of her son and the fear that Belle would join them soon.

Cassie's tale ended with Gideon vanishing into the mist, saying he was on his way to Leith.

'So he left you alone?' Flynt said, avoiding the boy's gaze again. It wasn't resentment in those eyes. It was hatred. And that pained him. 'Those men could have returned to complete the task.'

'For some nights I didn't sleep soundly, but we took it in turns to keep watch with pistols in our laps.' Cassie contemplated her next words. 'We also had assistance.'

'From whom?'

A pause, a long one, as she once again considered. 'A friend.' She diverted further questioning by opening the small bag attached to her wrist and producing a crumpled sheaf of papers. 'I learned of Gideon's plan in this letter, which he had delivered by messenger. It says much and little in similar measure.'

Flynt took the letter and recognised Gideon's expansive hand immediately. He unfolded the paper but was prevented from reading by a muted commotion at the door, which flew open to reveal Belle, still in her finery, with Jerome, the house guardian, behind her, shrugging apologetically at Flynt. Belle's eyes took in the mother, the boy and then, finally, Flynt.

'I understand we have guests, Jonas,' she said.

'We do,' he said, rising, the letter still in his hand. 'Belle, this is Cassie. My stepsister.' The latter was added after a pause, as if it might defuse any anger that could be forthcoming.

'Cassie,' Belle repeated, her scrutiny returning to the woman in question. 'I have heard much of you.'

She knew who she was, of course, for Flynt had murmured her name in his sleep, even cried it out some years before when reaching the climax of their lovemaking. He had ensured never to do that again, but Belle would remember, of that he was certain. Seeing them together now, discomfiting though it was, he was amazed at how alike they were. The same size, the same bearing, the same flawless skin, the same fire in their eyes that could sometimes be shrouded with sadness but was not slow to kindle. Both pairs of eyes now sized each other up, deciding whether this was friend or foe. Or rival.

'And this is her son,' he added, hoping to distract Belle further.

'Jonas,' added Cassie.

That made Belle shoot a brief glance in his direction. That glance told him he would have some explaining to do when they were alone. Nevertheless, she smiled. 'You are both most welcome. I am Belle St Clair, and I am joint owner of this house. Flynt did not inform me you were visiting.'

'He didn't know,' said Cassie.

A delicate frown creased Belle's forehead. 'There is something amiss?'

Belle's acuity was sometimes disconcerting. She could take a reading of a person's character within minutes, seconds, of meeting them. It was a skill she had picked up over the years working for Mother Grady, for she had to discern whether a cull was dangerous.

'A family matter,' Flynt said. 'My stepmother has been abducted and my father has taken up the pursuit.'

'Abducted by whom?'

'Slave catchers,' Cassie said.

Belle's eyes, until now shadowed by curiosity and good humour, hardened. She had been a slave, sold as a child on the block at Kingston and transported across the Atlantic to be schooled in the art of the courtesan. A slave she had remained, though treated reasonably well given her new owner's profession, and in some comfort, until Mother Mary Grady's hard heart, softened by age and affection, gave Belle her papers and even made her partner in the house.

She nodded, just once. 'You must be fatigued after your journey.'

Flynt stood. 'I will find Cassie and her son lodgings...'

'Nonsense,' Belle said. 'We have spare rooms here. You may rest.' She stopped, as if a thought had struck. 'Unless you have moral objections to the business we transact?'

Cassie held her gaze. 'Miss St Clair, I spent my early childhood with my mother in a tavern in Nassau. I saw morality in its rawest state. I suspect the business here is more civilised.' She returned her attention to Flynt. 'But it is not refuge I seek.'

Belle asked, 'Then what?'

'Passage to New Providence. I will find Gideon and my mother and will see them both returned safely.'

Flynt displayed no surprise at the response, for he had expected it. He looked towards young Jonas. 'And what of the boy?'

'I'm no boy!' This was the first time he had spoken, and the three words contained a considerable degree of heat, not through outrage at being dismissed as a boy but because of who had done so. That heat also warmed his cheeks. 'I am near eighteen.'

'I apologise,' Flynt said, his words sincere. 'I spoke without thought. You're a young man, of course, but I'd hazard not used to the rigours of such an enterprise.'

'What would you know of that to which I am or am not suited, sir?'

Flynt noted the contempt layered heavily on the word 'sir', but had no desire to argue with him. 'Again, my apologies. I have no reason to think you are not suited.'

Apart from the fact that you look like a scholar, he thought, *and have perhaps never raised a fist in anger in your life.*

Cassie sprang to her son's defence. 'Jonas was most energetic in my defence when those men forced their way in.'

Flynt inclined his head in submission. 'I understand. But the course of action you suggest is fraught with dangers. The voyage alone, especially this early in the year, will be perilous, and there will be further dangers once landfall is made.'

'I can look after myself,' young Jonas insisted.

Flynt let that lie and addressed Cassie. 'Gideon can also look after himself.'

'So you would allow him to face this alone?' Cassie accused. 'He is not the man he used to be.'

'Gideon Flynt is more of a man than any other I have met. He is most capable.'

Cassie stared at him with something akin to contempt. 'You will not help him? You will not help us?'

'Of course he will help you,' said Belle. 'After all, it's what Flynt does, is it not?'

He couldn't tell if she was mocking him or giving him her blessing. That was a matter to be discussed later in the evening.

When Flynt didn't answer, Cassie sensed something in his silence and her eyes narrowed in understanding. 'I know your intent. You would have us tucked away safely and then you will pursue him, is that it? You would simply vanish without a word? Don't deny it, for you have done such before.'

He couldn't deny it, because that thought had indeed crossed his mind. In his youth he had been both intemperate and impulsive. The love affair that had turned physical between he and his stepsister had not prevented him from running away to find adventure. He had found adventure but it was not what he had expected. It had been bloody and brutal and had revealed skills within him that he used to this day, skills that had kept him alive, if not others. What he had not known when he left was that she already carried his child.

He sighed. 'I cannot seek Gideon and worry about you and... your son.'

'You need not worry about me. I will look to myself, and I do not expect you to accompany me. I only come to you because Gideon wished it, but you are correct, such a journey is no place for Jonas...'

The young man's gaze switched to her and became accusatory. 'You would leave me?'

'I need you to be safe. You will remain here with your...'

She broke off suddenly and there was an uncomfortable silence in the room.

'With Flynt,' she completed, but everyone knew what she almost said.

The young man shot around the chair. 'I will not!'

'You will do as I say,' Cassie said, her voice unbending.

'I'll abscond the first chance I get.'

'I forbid you to do that.'

'I'm a grown man, Mother, and I love Grandfather as much as you. I will not remain here with this man.'

The last two words were once again spat out coated with contempt. Flynt could not blame him.

Cassie and her son glared at each other, both unyielding in their position. It was Belle who broke the silence. 'This is not a subject to be discussed when you are both exhausted from travel. You have come a long way and passions are high, so I suggest you rest a few hours and revisit the matter in the morning. Come, I shall take you to the spare chamber and make you comfortable.' She turned to Jerome, who lingered behind her. 'Be so kind as to have some hot water taken up. A bath, I believe, would be most refreshing.'

Flynt was amazed that Cassie acquiesced so easily and allowed herself to be ushered from the room. Young Jonas hung back, his youthful desire to remain in order to further vent his anger towards Flynt taking precedence over his desire for hot water and a warm bed. Finally he followed his mother. Before she closed the door, Belle gave Flynt a long look that he couldn't decipher.

Flynt stared at the door for a few moments, wondering at how life can change in an instant, and then opened his father's letter again.

3

Cassie, my child,

 I write this before I set sail for the Indies by way of explanation. I believe you will be safe for the time being but I beseech you to seek assistance. The men who came to your home were in the employ of another named Toby Hawke, a foul creature who you may recall from your childhood but one I thought dead, though it is now most clear that he is not, for I saw him with my own eyes in the High Street as he took your mother. I suspect he will return her to New Providence where she will once again be his property, in his eyes at least. I did not act swiftly enough to prevent the abduction and that is one of many regrets I will take to my grave, but not before I put him finally in his. He allowed me to hear that he was bound for Leith but I now know he was not; perhaps it was to Greenock. I did something similar when taking your mother and you away, to confound any pursuit. As far as they knew we left New Providence for the American colonies. Hawke was unaware of my true name, also that I was from Edinburgh. No matter, I have lost too much time and will not catch him now, not in Scotland, so that means I must take ship myself. I will work passage on a merchantman, bound on the morrow for the Indies, and will reach New Providence when I am able. I pray to God that I reach her in time, for I fear Hawke means to punish her for my crimes.

 As for you, my darling Cassie, you must look to yourself, for he may have agents yet who will do their best to take you. I know you will have some assistance from this man who has become close to you but I urge you to put aside your differences and seek Jonas out. He wronged you grievously, but that is in the past and you moved on to a new life, a better one with Rab than he would have given you. You know that to be true. He has kept himself distant from you and young Jonas because that was what you wished, but he has not kept himself from me, not of late anyway. I've had letters, only a few, but that is more than he wrote during those years we thought him dead on some foreign field. Seek him in London, at the house of a Mrs Mary Grady in Covent Garden, and if *he be not there then they will know where to find him. Be not shocked by the manner of house it is and do not judge. Seek him out and place yourself under his protection. Yes, I hear you even at this distance insisting you can look after yourself, and I know that to be true, for you have learned well the use of pistol and even blade. But even so, you will be no match for the kind of man Hawke might send. Leave Edinburgh, take the coach or a ship to London.*

Jonas, if you read this then Cassie has done as I suggest. You will have gathered from the contents of this missive that I have not been fully open with you regarding my years at sea. You thought me to be a merchant seaman but in truth I was pirate. I didn't use my given name, because you and your mother were at home believing me to be an honest seaman, and if we were ever brought about and I faced trial then I didn't wish you to suffer any shame. They knew me as Sam Bell, and thought me from Glasgow by way of the Virginia colony. Under that name I was party to many acts of which I am not proud, something else I feel sure you can understand.

Jonas, you must take care of both Cassie and young Jonas. They are your responsibility. Do not be tempted to follow me for I can shift for myself most expertly.

Be safe, beloved Cassie, and with good fortune and God on my side, your mother and I will be back with you within a twelvemonth.

Yours, with love,
Gideon

4

The musicians had fallen silent. The revels had ended. Flynt heard voices in the foyer, goodbyes being made, the front door opening, closing, carriages clattering on the Piazza outside, but made no move to play the host and make his own farewells. He sat alone in the small reception room, the letter dangling from his hand, his mind upon its contents.

The revelation that Gideon had left the man Hawke for dead was no surprise; he had told him as much when he had last been in Edinburgh. Equally, the statement that his life at sea had not been entirely legitimate came as no shock. Flynt had long suspected such, given the size of his father's purse when he finally returned home with Mercy and Cassie in tow, there being sufficient funds to purchase the tavern and keep them all in relative comfort. His ease with weapons, which he had passed on to both him and Cassie, suggested more than a passing facility. Gideon was a good man, a kind man, and yet Cassie was correct – he was older now, and this pursuit could prove perilous for him. Flynt knew his temper well because he had often tested it as a youth and even now that he was full-grown he was loath to draw his ire, but he would have to ignore the stricture against following him. How to accomplish that and convince Cassie, and her son, to remain in safety in London he did not know. Even if he did manage such a feat, Cassie was strong of will; he would have to make arrangements for their protection.

And then there was this friend with whom she had grown close. He had sensed an evasion from her in that regard. Who was he? Why was she so circumspect? And what was the meaning of the sharp pain that pierced his chest and the tightening of his throat at the thought of them being close?

Belle reappeared, closing the door firmly behind her and, without uttering a word, moved to the chair in which Cassie had been sitting, placed her hand on one of the wings and regarded Flynt.

'Cassie is a beautiful woman,' she said.

'That she is.' His voice was gravelly, so he cleared it a little with a cough.

'I can see why you love her.'

'I did, at one time.'

He had believed that to be true. At one time. Now he was unsure.

'But not simply as your stepsister,' Belle said. 'She *is* the one whose heart you broke all those years ago, yes?'

Flynt had told her only a little of Cassie after breathing her name. He gave her a brief nod.

'And she married another?'

'A boyhood friend, Rab Gow.'

'And where is he?'

'Dead.'

She paused. 'By your hand?'

'No.'

That wasn't strictly true. He had not pulled the trigger but he had caused his friend's death. At least, that's the way he saw it. Cassie might see it differently, though she had once agreed that Rab's own actions paved the way to his end.

There's a darkness that follows you...

A conversation he'd once had with her echoed in his mind.

I fear you carry the stench of death...

'The boy favours you,' Belle said.

He was brought sharply out of his memories. 'You know?'

Her smile was limp. 'Anyone who observes you together would see it. Does he know?'

'I believe so.'

'And you knew of him?'

Why did you leave without a word? Why were you so unfeeling?

Cassie's words again. He took a deep breath. 'I learned of it when I returned to Edinburgh three years ago.'

'Cassie had kept his existence from you?'

'I had left Edinburgh as a youth seeking adventure before her pregnancy was apparent. I failed to communicate after that, and they believed me dead.'

'And you have known these three years since and never mentioned him.'

You were cruel…

He couldn't find the words to explain himself. Belle's stiff expression told him she knew he would not.

'This friend of yours, the one she wed, did he know?'

He swallowed. 'I think probably yes.'

'He was a good man then.'

'He was.'

'And your father, he is a good man?'

Flynt thought about the contents of the letter in which he admitted he had committed acts of which he was not proud. Flynt could not judge, for he had his own shame and regret. 'He is.'

Belle rose. 'Then for the sake of Cassie and her son, and your father, you must help them.'

'I would be absent for perhaps a year, maybe even more.'

She was already at the door. 'Then that is the way it must be.' As she stepped from the room, she added, 'Colonel Charters awaits you in the salon.'

There was something in her voice, a slight hitch to the words, perhaps an emotion she kept down trying to break through. She swept across the foyer towards the stairs as if she had something pressing to which she must attend.

5

Charters was alone and seated in one of the armchairs near to the fire, which blazed, a book he had taken from the shelf open on his lap as he sipped a glass of wine. Flynt needed a drink, too.

'You have been absent from your own festivities, Serjeant,' Charters said as Flynt turned gratefully to a small table behind the door carrying the bottles of wine, port and brandy.

'I had visitors,' Flynt said, pouring himself a liberal measure of brandy and swallowing half of it, before turning to face the man who exercised considerable control over his life.

'Not welcome visitors, I'll be bound, given the way you quaffed that brandy.'

Flynt stood over the fire, one hand resting upon the high mantle, and stared into the flames. 'Awkward, more like.'

Charters remained silent, his eyes falling back to the book. Flynt knew this to be a pose. He was expected to amplify his remarks.

'It was my stepsister and her son from Edinburgh; I'm sure you know of them.'

Charters made a point of knowing everything there was to know about the men and women he recruited to his Company of Rogues, an organisation made up of individuals who were thieves, gamblers, vagabonds and killers. Men like Flynt. The company was born from the domestic spy network set up first by Sir Francis Walsingham during the reign of Elizabeth, further honed by Sir William Cecil and his son Robert before, several decades having passed, resting in the capable care of Nathaniel Charters.

The colonel did not dispute his knowledge of Flynt's family. Not for the first time Flynt pondered how much he knew. Did he know of young Jonas, for instance?

'As you will be aware, my father returned from the sea with a new wife and her child.'

Charters made no response as he continued to read his book. He had not turned a page, however, for to achieve such he would have to put his wine glass down. Charters had only one arm; the other he had lost during the Battle of Malplaquet.

'He had taken that wife from a slave owner, who has now tracked him down and taken her back,' Flynt continued. 'My father has set off for the Indies to retrieve her.'

The glass was now set down on the floor beside the chair leg, the book was closed and Charters looked up at him. 'And you wish my permission to go after him.'

'I don't need your permission.'

'That is open to some debate. At any rate, it is not given.'

'As I said, I don't need…'

Charters' voice sharpened. 'I heard what you said, Serjeant, and we both know that you are bound to me by guilt; do not forget that.'

A charge of theft and assault was held over Flynt's head like the sword of Damocles to force him to work for the Company of Rogues.

Flynt controlled his temper. He'd had a trying evening, and it wasn't easy but he needed Charters on his side. 'I have to go after him.'

'I have work for you here.'

'You owe me a debt.'

'Really? How come you to such a surprising conclusion?'

'I saved your life by dragging you from the field at Malplaquet.'

'I save yours every day by not having you arrested for robbing the duchess and her young swain.'

'I didn't do that, and you know it.'

'It would be interesting to see what position the courts would take on that point.'

They had argued this many times, and each time Flynt became more convinced that he was well and truly trapped. Even though he knew who had actually robbed the duchess's coach and beaten her lover senseless, there was no way out. Charters would easily produce bogus evidence. A false identification, a witness who was never there. If he chose to have him arrested and tried, his life would be forfeit.

'I have performed many tasks for you, some of which would bear little scrutiny.'

'You did these things for the good of the nation and have been well compensated for your labours. And you cannot prove any connection to me.'

Flynt drained what was left of his brandy. 'I helped you find some peace of mind regarding Nimrod Boone.'

That brought a response, a twitch of the eyebrow, nothing more, but it was proof that perhaps this thrust had hit home. Nimrod Boone had been a man who had haunted Charters' mind for years until Flynt managed to engineer justice.

'That was a personal matter,' Flynt pressed.

'That was still a matter of security,' Charters said. 'Boone was a murderer and threatened the peace of the city.'

'You may justify it how you wish, but the fact remains that I did the work for you as a favour.'

Charters permitted himself a small smile. 'We both know that was not the case, Serjeant. As I recall, you were first placed on the trail on behalf of one of your criminal contacts.'

'His interests intersected with yours. You allowed Boone to escape in Flanders. People died because of your inability to hold him. Innocent people, Colonel. Children.'

Charters flinched as if some sharp pain had pierced him, and changed tack. 'We have business outstanding with the Fellowship. They continue in their endeavours to control our society in order to enrich themselves.'

'You have others within the Company of Rogues who can deal with them.'

'None like you. None with your experience of their methods and schemes.'

On any other occasion, Flynt would have been flattered, but he believed his encounters with the shadowy group of businessmen and criminals were of minor irritation to them. 'I have barely drawn blood and you know it.'

'You are a thorn in their side that I would rather pricked them further here than have you galloping across the world. No, Serjeant, I refuse you permission to fly off on some family quest. The Caribbean is a large sea and you are no sailor. You cannot hope to find him.'

'I know where he is headed – the island of New Providence. That will be a start, at least.'

Hearing the destination caused Charters to frown. 'To Nassau?'

Flynt felt something shift in the air between them. 'Aye. The man Gideon seeks owns a tavern there.'

'And what is this man's name?'

'Toby Hawke.'

Charters mulled this information for a moment but then shook his head. 'You are on a fool's errand. You cannot hope to find him and, besides, neither you nor your father has a case. Like it or not, the woman was the legal property of this man Hawke and remains as such. Your father committed theft when he took her.'

'I don't think Gideon will turn to law.'

Charters pursed his lips. 'Like father, like son. Nassau is a cesspit of villainy. He will find himself outnumbered, I fear.'

'All the more reason why I should go to his aid.'

Charters considered again. 'And if I continue to refuse my permission?'

'As I said, I do not ask permission, I am merely informing you of my intentions.'

A sigh from the colonel. 'You are truly exasperating, Flynt, have I ever told you that?'

'Many times.'

'And yet you still find new ways to exasperate me. How do you do that?'

'It's a talent.'

Charters picked up his wine glass and drank what was left. 'Attend me tomorrow, at an hour after noon, in the Black Lion.'

'You will not deflect me from this path, Colonel,' Flynt said, his voice sharp. 'I will go, whether it is to your liking or no.'

Another sigh. 'So exasperating. See me on the morrow. I will study upon it overnight.'

6

Flynt had intended the evening to end in a far different way than returning to his lodging in the Golden Cross at Charing Cross. His plan had been to finish what should have been a delightful occasion in Belle's bed. But life, as he had remarked before, often made mockery of a man's plans, and so he donned his greatcoat, his hat, his pistols Tact and Diplomacy and silver cane, and took to the dirty, cold streets. Rain had drenched the city for days on end, so torrential at times that he wondered if someone, somewhere was building an ark, but it had dissipated, though the night still held the promise, or threat, of more. Nevertheless he walked, in the hope that the brisk air would not only overpower the stench of London but also relieve that throbbing in his temple that had built over the course of the evening.

He left the Garden via Henrietta Street and thence down Bedford Street to Half Moon Passage and thence to the Strand. It was while heading west that he became aware of the two men shadowing him. It was late, but Londoners still walked the streets. Doxies looking for culls, culls looking for doxies, gentlemen moving from one tavern to another, brothel to gaming house, coffee house, Molly house, eating house. Carters trotting their horses to and from places of business. Thieves of all sorts in search of prey. Divers alert for a carelessly disregarded pocket from which to lift a silk kerchief or money pouch. Cutpurses and footpads lurking in the shadows. Link boys holding aloft lanterns or torches to light the way, sometimes for legitimate reasons, sometimes to lead the unwary to a gang that lay in wait in some dark lane or alleyway. This was his London, a warren of streets pregnant with humanity in all its forms: high and low, honest and crooked, moral and immoral. In his part of this world there were more of the dregs than the cream of society, and he suspected the two men who kept their distance were not following him to discuss philosophy. The way they held themselves, the way they walked, the way they kept him in their eyeline while also being aware of their surroundings told him they were professionals. He couldn't see their faces, for the few lamps and lanterns burning outside doorways were near to ineffectual, so could not tell if he had previously had their acquaintance and they were perhaps seeking some form of redress.

He had kept his coat unfastened, and as he walked he positioned the flaps in such a way that, should it be necessary, he could reach his pistols with ease. He also twisted the handle of his cane to loosen the blade within. If they came at him, and his life was such that it was a near certainty, he was ready.

He was almost at Charing Cross. The coaching inn that was currently his home was on the right, and a carriage lingered in front of it, a coachman and driver waiting. It was then that the men behind him diverged, one remaining at a distance, the other quickening his pace. They were indeed professionals, for if Flynt took one, the other would easily pick him off. Even so, he whirled, sliding the thin blade free from its silver sheath, but the man halted a good four feet from him, both hands visible and raised in a placating manner.

'Easy, friend,' he said, his accent not London. Somewhere in the north, perhaps, but not as far north as Scotland. 'We mean you no harm.'

A swift glance at the second man showed Flynt that no weapon had been drawn. 'What do you want then?'

'We're here only to convey you to that coach yonder.'

'For what purpose?'

'Our employer wishes to have words.'

'Who is your employer?'

The man waved a hand. 'If you would be so good as to step to the coach you will see for yourself.'

'And if I decline? What then?'

Both hands were raised, palms upwards. 'Then it's no skin off my back, friend. We have no further instructions than to bring you here, but as you were headed in this direction anyway, we kept our distance.'

Flynt detected no equivocation in his manner. The man spoke the truth, so he eased the blade into its sheath but did not lock it. He didn't know who he was going to see but experience had taught him it was best to be prepared for the worst in any invitation, even when the one extending it seemed more fair than foul.

He crossed the final few feet to the waiting coach where the coachman had swung down and already held the door open. Flynt ensured there was a safe distance between him and the door as he peered within, seeing nothing at first but a pair of legs in elegant hose and shoes, and fingers resting on a cane with a silver handle in the shape of a wolf's head. A face loomed from the shadows and he took a further step back, his blade half bared.

'Easy, Flynt,' said Lord James Moncrieff. 'I come as a friend.'

'I don't recall us being friends,' said Flynt, giving the coachman a quick study but seeing no threat. The two men who had shadowed him were now on the other side of the Strand, displaying no sign of aggression.

'I assure you I mean you no harm,' insisted Lord Moncrieff.

That would be something new, Flynt thought. This was the man who had engaged a hired killer to stalk him, then lure him to a village in the north, where a mad nobleman had already dug his grave. This was the man who was a ranking member of the Fellowship. This was also the man who was his half-brother, a fact that had come to light in recent years after Flynt was told that Lord Moncrieff the elder had raped his mother in Edinburgh. The fear that Flynt would try to claim some sort of inheritance had increased the enmity caused by Flynt's efforts against the Fellowship.

'What is it you want?'

'First I'd like you to step inside and permit us to discuss our business in private.'

Nothing on this earth would convince Flynt to step into that coach. 'Thank you, but I'm more comfortable here.'

'The night air is cold and dank.'

'I find it invigorating.' When Moncrieff let loose an irritated sigh, Flynt couldn't help but smile. 'I've been told I can be most exasperating.'

'That you can. I am most sincere when I say that I mean you no harm, on this occasion.'

'And I believe you, but you can say what you came to say from there.'

Moncrieff's eyes flicked to the coachman who still held the door open. 'Thank you, George, you may join our friends across the street. I'll call you when I am ready to leave.'

The coachman inclined his head and did as he was told.

'There, now that we are, to all intents and purposes, alone, what can I do for you, Moncrieff?' Flynt asked.

'I come about Gideon.'

Flynt was instantly on the alert again. 'What about him?'

'You know what took place in Edinburgh?'

'I do.'

'You know he has gone after the men who took the woman, Mercy?'

'His wife, Mercy,' Flynt corrected. 'My stepmother.'

'Quite so. Do you know the manner of man he faces?'

'Toby Hawke.'

Moncrieff was surprised that he knew the name. 'You have heard of him?'

'I heard the name this very evening for the first time.'

'He is a most dangerous individual.'

'I take it you are acquainted with him?'

'We have had business.'

'Fellowship business?'

Moncrieff shrugged that away. 'I met him in Edinburgh to discuss mutual interests in the colonies. He asked about the man who owned Gideon's tavern – apparently he'd seen him walking in the street one day, saying that he believed he'd known him in Virginia. I thought nothing of it.'

'But you discussed Gideon with him? And Cassie? Where she lived?'

'He seemed to only have a passing interest and, as I said, I thought nothing further of it. And then I heard what had occurred. We have had our differences, you and I, and no doubt will again, but I mean your family no harm, Flynt, you must believe that.'

Despite his growing anger, Flynt did believe that. Moncrieff was a criminal, sweet-smelling, finely tailored but as much a rogue as those who lay in wait for unsuspecting victims. The Fellowship kept itself to the shadows but it was involved in many murky schemes.

'Has Hawke returned to New Providence?'

'That was why I wished to speak with you. I understand Gideon has taken ship for the Indies and plans to find him there.'

'How do you know that?'

'It is the talk of Edinburgh, among certain circles at least. I have eyes and ears and lips that see and hear and tell.'

He meant the Fellowship's eyes and ears and lips, but Flynt didn't interrupt.

'Hawke is waiting for him. He means to repay Gideon for the ill he inflicted upon him when he stole the woman and child away.'

'Mercy and Cassie. They have names.'

Moncrieff held a hand up in apology. 'Of course. Hawke is not a man to be trifled with, and Gideon is no longer young.'

'Why do you tell me this?'

Moncrieff took his time in answering, as if he was trying to understand the reason himself. 'Because you spared my life in the north and I don't like being in another man's debt.'

'Especially me.'

'Especially you,' Moncrieff agreed.

'You inferred at the time that what I did changed nothing between us. Why the change of heart?'

'I said that if you crossed the Fellowship it would not prevent me from doing what I needed to do.'

'It would seem this man Hawke is part of your Fellowship. If I go after him, if Gideon goes after him, then are we not both crossing you?'

'I didn't say he was part of the great endeavour,' Moncrieff said as he settled back into the darkness of the coach. 'But if he were, sometimes sacrifices have to be made…'

7

Flynt slept fitfully, his mind a whirl of fears for Gideon, of Cassie returning to his life, of the identity of the man who had been her protector, of Belle's sense of betrayal, of young Jonas's clear distaste for him, of Moncrieff's intercession and the ease with which he seemed to discard Toby Hawke. What could that mean? What little Flynt knew of the Fellowship suggested that they were protective of their anonymity, although Charters had in recent weeks discovered the name of its Grand Master, a man called Andrew Wilson, who on the surface was an obsequious Edinburgh council official. He had protected his true character well. Flynt had met him, had even on one occasion had cause to intimidate him to obtain information, but so artfully had the man played his role that even he had been taken in by his alter ego.

The question was, why would Moncrieff, himself a ranking member, betray one of the basic rules of the Fellowship? Especially to Flynt, who he hated. The suggestion that it was out of concern for Gideon didn't carry water. Moncrieff's family cared nothing for anyone but themselves. No, he had his own reason for steering Flynt towards the man Hawke, but such was the circuitous route of Moncrieff's mind it was unlikely his true motive would be revealed until he was ready. If ever.

Flynt rose early, washed himself down with flannel and soap, cleaned his teeth with powdered oyster and eggshells, breakfasted on coffee and bread with milk, courtesy of his landlady, Mrs Wilkes, then called upon the stables where he housed Horse. He had to travel to Wapping, and if he was to keep his appointment with Charters, then he must make haste. He had the boy saddle up the mare while he spoke to the stable owner, a retired officer of the infantry who had always wished to join the cavalry. Flynt informed him that he may be away for some considerable period but would leave ample funds to cover the horse's care. The man gave his assurance that she would be well looked after and repeated a previous guarantee that nobody would try to ride her, for that way could lead to broken bones. Horse was intensely loyal to Flynt and would allow no other person to mount her.

It felt good to be back in the saddle, especially on a day that was shaping up to be dry, if cold, though the skies glowered darkly. He trotted Horse at a steady pace through the still-dark streets to the Strand, then on to Temple Bar, passing under Mr Wren's fine archway to Fleet Street and over the bridge to the Ludgate before cutting down towards the river, passing the Tower and onwards to Wapping. The journey took him nigh on an hour and a half, thanks to the volume of carts, chairmen and carriages as well as flocks of sheep, geese and other livestock being pushed towards the various markets. The light began to rise as he arrived at the sign of the Ship and Anchor close to Execution Dock, and Flynt was glad to see the iron-grey skies slashed by slivers of blue. Were it not so early in the year he would have sworn there was a hint of spring in the air.

He found a small stabling area at the rear of the inn and paid the lad there a sixpence to give Horse a rubdown and, in a little while, some water and feed, then made his way back to a side door and climbed a steep set of stairs to the office used by a man he knew only as the Admiral. His ever-alert guard, Daniel Pickett, no doubt having heard his footfalls on the steps, met him at the top, his face displaying his customary lack of expression. He held out his hands and Flynt gave up his pistols and cane. There were few for whom he would do this but he trusted these men. A nod towards the door instructed him to enter, where he found the Admiral behind the desk framed by a window looking out on to the river. The floor and ceiling of this room were set at curious angles so that Flynt believed himself to be walking uphill.

'Jonas, my good friend,' said the Admiral, his voice rheumy, his face, as ever, concealed behind a leather mask, with only a slit for the mouth, twin perforations for the importation of oxygen and one single eyehole. He didn't rise from his chair, nor did he extend either of his gloved hands. Flynt took no offence. The extensive burns the man had received during a naval engagement precluded such niceties.

'You wish something from me?'

The early hour, the fact that this meeting was not planned, would have told the Admiral that Flynt had a purpose. He oversaw the criminal activities on London's waterfront and riverbanks and was no fool. He was ruthless, he was cunning, and yet Flynt liked him.

'I need a ship,' said Flynt, getting straight to the point.

A soft laugh soaked with liquidity seeped from under the mask. 'Who do you wish to smuggle from the country this time, my friend? Not another Scottish noble whose head is destined for the axe, surely?'

Flynt had previously required passage for such a nobleman and his wife and had prevailed upon the Admiral to provide passage to the continent. 'No, this time it's for me.'

'For you?' The ragged voice was surprised. 'You must flee the country? What have you done? Have you misbehaved in some fashion that requires a swift exeunt, as the theatricals might say?'

Flynt smiled. 'Always, but I'm not avoiding the authorities. I must reach the Indies quickly.'

'The Indies is a godforsaken hellhole of heat and insects and individuals of most peculiar morals.' Another wet chuckle, for they both knew London was replete with individuals of most peculiar morals. 'Why would you wish to go there?'

Flynt told him the truth. Of Gideon and Mercy, of Cassie and young Jonas. Of Toby Hawke. When he was finished, the Admiral nodded. 'Then of course I must help you. After all, did you not risk all to avenge my poor departed friends?'

It had been the Admiral who first set him on the trail of a murderer who was later revealed to be Nimrod Boone. Flynt damn near lost his own life in bringing him to justice. A summary justice, it turned out to be.

'Do you have a vessel embarking for the Indies?'

'I have two in my fleet, leaving these shores right soon.'

'You have a fleet?' Flynt couldn't hide his surprise.

Another wet chuckle came from the leather mask. 'I own a number of vessels, my friend, through intermediaries, of course. You will travel alone?'

During his disturbed night, Flynt had come to the realisation that there was little chance of preventing Cassie from making the journey, and Flynt saw enough of himself in the boy to convince him that he would fight tooth and nail to accompany them. 'There may well be three.'

'Your sister and her son, I presume?'

Flynt nodded, expecting a protest but receiving none.

'How soon can you be ready?'

'More or less immediately,' Flynt replied.

The Admiral grunted in satisfaction. 'You may recall there was a sea captain in this very room of recent, making report to me. I told you he plied the trade routes between here and the colonies for me. His vessel, the *Sprite*, will set sail on the morning tide, carrying timber and cloth to Jamaica and returning with sugar. I would be most keen for you to be able to sail with it. There is a cabin for such, though comforts be few, as long as you all don't mind bunking together.'

Cassie would object to such an arrangement. 'I could perhaps find a corner in the crew's quarters.'

The Admiral shrugged. 'It can be arranged. The men who make up the crew are not what you might call polite, but you are no innocent and I reckon you can endure the rigours of rough talk and rougher living.'

Flynt's knowledge of the geography of the West Indies was far from adequate. 'How close is Jamaica to New Providence?'

'Is that where this rogue Hawke has taken the lady? Of all the hellholes in the Caribbean, that is by far the worst.'

'Nevertheless, that's where my father has gone and that's where I must go. Is there much distance between the two islands?'

The Admiral made a calculation. 'Over one hundred and thirty leagues, a distance of around four hundred nautical miles.'

'If your captain…'

'Johnson, Captain Charles Johnson,' the Admiral said. 'You will like him, for he is a man of letters, most well-read and with a desire to pen a volume or two of his own, which be most unusual for a gentleman of the waves, who are more prone to writing in a logbook than anything more creative.' He paused to take a ragged breath. 'He is an honest man, carrying an honest cargo.'

Flynt understood that the Admiral was making it clear that this particular financial endeavour of his was legitimate.

'If this Captain Johnson were to convey me to Jamaica, would I find a vessel there that would take me to New Providence?'

'Anything is possible if the right price is met. But I would urge you to have a care, my friend.' The Admiral's voice had turned serious. 'I would caution you against this adventure. The Atlantic can be capricious and there will be many hazards. Do you have the gut for the journey?'

Flynt was not a natural sailor, but when crossing to the continent during his military service he had not fallen foul to the vomiting plague at sea and had managed to retain his stomach, unlike many of his comrades. He recognised that the Atlantic was not the Channel, however. 'I believe I can endure it.'

'That is to be hoped, for I have seen men turn green when faced with waves as high as the walls of the Tower. But that is not the only danger you face – the worst will come when you make landfall. You will have to have your wits about you, Jonas, for there are more cutthroats in that part of the world than in any other. It is a lawless place.'

'More lawless than London?'

The Admiral laughed again, this time with an added wheeze. 'London is the Garden of Eden in comparison.'

'Worry not,' said Flynt. 'As you say, I'm no innocent. But there is one further thing you might do for me, if you will and if you can in the time available.'

The Admiral spread out his gloved hands. 'Name it.'

Flynt named it.

—

Flynt took his time on the return journey to the city, savouring the dry weather, taking in the sights of the city streets, enjoying being with Horse, even stopping by a street vendor to purchase some apples. They had been through much, animal and man. She had been with him during his sojourn on the heaths as he lived the life of the highwayman, earning him the honorary title Captain, as such men often are. Road piracy, they called it, so perhaps that required him to carry some sort of rank. During his military career Flynt had only reached as far as Serjeant and that for only a short time before his innate antagonism for authority got the better of him. That he had been so promoted was solely down to his rescuing the wounded Colonel Charters. Why the officer had been on that Flanders field in the first place he had discovered only recently, it being all tied into horrors that the man Nimrod Boone had perpetrated during his army service and continued more recently in London. Flynt himself had been fleeing the carnage of the battle when he chanced upon the colonel, near death, his arm all but hacked from his body by Boone's sabre blow. He didn't know if Charters was aware that Flynt had been in the process of deserting, but it wouldn't surprise him. The man was unnervingly prescient in most matters.

Back in the stable, once Horse was unburdened of saddle and treated to an apple, sliced and the core discarded, Flynt lingered, caressing her ears fondly. Most mornings, unless he was on a mission, he would come by and spend time with her. Sometimes they would ride, other times he would walk her. She would be well taken care of, he knew that. But still, he would miss her. Her strength, her stamina, her speed, her loyalty, had helped him escape many a tough situation. There were times he thought himself incapable of love, Belle's words about requiring a heart for such an emotion hitting home, but he loved this animal.

As he stood alone in the stable, his hands still rubbing her ears, his forehead resting on her muzzle, he muttered the words he'd never been able to say to another person. He liked to think she understood.

8

The Black Lion was filled with post-midday revellers, which differed from the pre-midday revellers only by the amount of gin, rum and ale they swallowed, the oysters and meat and bread they ate and the tobacco they smoked. The vestiges of the latter filled the air as Flynt entered. He had never taken much to the pipe, but he didn't mind the aroma of the burning weed, if only because it helped counteract the stench of the tallow candles that had to burn night and day in the tavern, for even in summer's brightness it remained dingy.

Jack Sheppard sat at a table, a tankard before him. His usual sprightly nature had clearly deserted him this day. Flynt didn't deviate from his path in order to engage with him as he headed for the doorway that led to the private room above. There was a time when he would have, but that time was dead and buried. The young thief had thrown in his lot with Jonathan Wild, the man who proclaimed himself to be Thieftaker General. Flynt rubbed shoulders with many men and women who walked the crooked path, the Admiral being a case in point, but the man in the mask knew what he was and made no apologies for it. Wild adopted the air of the honest man, of one in service to the city, but he was as base a villain as any street thumper. Flynt had warned young Jack about joining his crew, but to no avail. Still, Flynt felt something heavy settle in his chest as he climbed the stairs. He liked the boy, had used his talents as a nimble-fingered thief and his incredible knowledge of the city's substrata on many an occasion, but had also encouraged him to follow the life of an honest man, even finding him an apprenticeship with a respected carpenter and locksmith. Jack had walked away from it all, and there was nothing Flynt could do now to help him.

Charters awaited him in the private room, the use of which the landlord, Joseph Hines, most readily granted whenever required. The colonel stood with his back to the large fire, which Hines had ensured was blazing to ward off the winter cold, his single hand behind him to warm it.

'Ah, Flynt,' said Charters, motioning him into the room. 'Close the door, there's a good fellow. There be a frightful draught a-blowing up them stairs.'

Flynt did as he was told and dropped his hat on the long table that filled the centre of the room. He kept his greatcoat on, however, because though he had enjoyed the milder feel in the streets, the air in this room was sharp and he suspected the fire had not been kindled long.

He decided to get right to the point. 'Do you agree that I must leave for the Indies? For I assure you, Colonel, that my mind is set upon this path and I'll not be dissuaded.'

Charters tutted. 'As blunt and argumentative as ever, Serjeant.'

'I've little time for social niceties. Every minute spent in idle debate is a minute lost.'

A smile tickled Charters' lips as Flynt became aware of a movement behind the screen that hid the area for men's necessaries. A tall man stepped out, drying his hands on a cloth. His bearing was erect, his body full but with muscle, not good living, his skin carrying sign of having been burnished by the sun, his cheek marked by a severe scar. His dark hair was long and tied back. His clothes were of decent quality and he gazed at Flynt with the eyes of a man who was used to giving orders.

'Serjeant Jonas Flynt, may I present Captain Woodes Rogers,' said Charters.

Flynt gave the man a brief nod, but didn't receive one in reply.

'Nat here informs me you intend to travel to New Providence,' Rogers said, taking a seat and settling back, one arm resting on the back of the chair, the other upon the top of the table. Flynt didn't know who he was but this was a man most comfortable wherever he found himself. His rank of Captain suggested both seafarer and army officer in equal measure, because it was doubtful he had been a gentleman of the road.

'I do,' said Flynt.

'I shall be voyaging there myself soon.'

So, he was a sea captain. Flynt was puzzled as to what business this man had with Charters. 'I wish you a safe journey then.'

'Captain Rogers has just recently been appointed Captain General and Governor in Chief of the Bahamas,' Charters explained. 'If you are so confoundingly intent on this quest of yours, then you can do him, and your king, a service.'

So Charters had acceded to Flynt's wishes. There had been a gleam in his eye when he discovered that Flynt would be travelling to New Providence; now he knew why. Charters' mind had already turned to this man.

'Nat informs me that you are a man of exceptional intrepidity, accustomed to thinking upon his feet in difficult circumstances,' said Rogers.

Flynt raised a single eyebrow towards Charters, who had never before praised him to another as far as he was aware. Charters rolled his eyes. 'Don't preen, Flynt, it's most unbecoming.'

'He also tells me that you are extremely comfortable with thieves and killers.'

Flynt couldn't deny that.

'I am in need of such a man,' Rogers continued, 'and the fact that you are bound for New Providence has proved decidedly *providential*, shall we say?'

Rogers and Charters both smiled at the little jest. Flynt remained stone-faced. 'What would you have me do? I'm sure Nat here told you I have business of my own to which I must attend.'

Charters glared at him for not only using his Christian name but also contracting it, just as Rogers had done. Flynt believed in taking his pleasures when he could, and anyway, he sensed he had something of a hand to play in this little game.

'He did, and I assure you that you will combine both tasks with ease,' Rogers said smoothly, then waved towards the chair opposite him at the table. 'Please, Serjeant, do sit. I get a crick in my neck somewhat woeful looking up at you.'

'Yes, damn it, man,' Charters said, considerably more tersely. 'No need to stand upon ceremony, we're all friends.'

Flynt was near to erupting in laughter. First Charters had sung his praises, and now they were friends. They really did want him to perform whatever tasks they had in mind. Flynt placed his cane beside his hat, followed by his two pistols. Rogers watched with some interest, his eyes sparkling with humour.

'You like your weaponry, I see,' he said.

Flynt pulled the chair from under the table and sat down. 'Being intrepid can be a dangerous occupation.' A small, but harsh, exhalation from Charters finally made Flynt smile a little. 'What is the nature of the work you require me to do?'

Before Rogers could reply, there was a discreet knock at the door and Joseph Hines himself appeared carrying a tray with wine, glasses, bread and cheese. Charters thanked him as he set them on the table.

'You will ensure we are not disturbed, Mr Hines,' Charters said.

'That I will, Colonel,' said Hines, his tone more respectful than normal. 'Your men has eyes upon the door below and the one to the rear. They shall not let no one pass.'

When crossing the tavern floor, Flynt hadn't played his usual game of picking out the men who watched Charters' back. They were seldom noticeable in such a setting and he enjoyed trying to identify them, but seeing young Jack so morose and alone had diverted him from the sport.

'I will, of course, also maintain a close watch,' Hines said. 'You is most secure here in the Black Lion, as ever, Colonel.'

Hines all but bowed as he left.

'You trust that man, Nat?' Woodes asked.

Charters was already pouring three glasses of wine. 'I do.'

With those words, Flynt became convinced that the tavern owner was a part of the Company of Rogues. Charters not only valued men who could take action but also those who could gather intelligence, and there was many a truth in rumour and many a secret that slipped when the tongue of the one holding it was loosened by liquor or the lubricious promise of a lady. The man behind the bar, or walking between the tables, was made all but invisible by dint of his ubiquity and that meant he was ideally placed to listen, to learn and to pass along. Charters was of the belief that if money was the power behind the spin of the world, then information was the hand that pushed it.

Flynt accepted the glass of wine but didn't drink from it. Rogers sipped his and leaned over to cut a hefty slice from the white bread, for Hines ensured there was nothing but the best for Colonel Charters and his guest. The fact that Flynt could also enjoy the more expensive loaf was an unfortunate by-product, he was sure. Rogers slapped a slab of cheese on the bread and bit into it, while Charters was more delicate in his approach, ignoring the bread but helping himself to the cheese and cutting it into smaller portions.

'Cheddar,' said Rogers after swallowing. 'Most fine.'

'The best cheese that England affords,' agreed Charters.

Flynt watched them eat before repeating, 'What would you have me do on New Providence?'

'Ah, yes,' said Rogers, putting his bread and cheese down, then following up with another sip of wine. 'To business. As Nat here has informed you, I am appointed Governor of the Bahamas and will take ship in two or three months to take up my post.'

'Why so long a delay?'

'I have much to prepare before I go. I will not be voyaging alone, for I must take a considerable force of men, as well as colonists and supplies, so must engage seven vessels.'

'Why such a large host?'

'Pirates, Serjeant,' said Charters. 'The scourge of the seas.'

'Exactly. These villains are the reason I have been appointed. Their depredations have a deleterious effect on England's trade, and the merchants both here and in the Americas, and the Governor of Virginia in particular, have demanded that something be done about them. And can you guess which island forms the heart of the pirate empire in the Indies?'

'New Providence, I assume.'

'Yes, indeed. Nassau, the principal settlement, is a nest of these vipers and it is my intention to flush them out. That is the task I have been set and, by God, I will perform it with the utmost efficiency. And, if necessary, I will be ruthless in my endeavours with those men who do not accept His Majesty's offer of clemency.'

'They've been offered clemency?'

Charters butted in. 'By God, Flynt, do you not keep up with affairs of a current nature?'

'I'm too busy being intrepid to enjoy the leisure of maintaining a working knowledge of events beyond London, Colonel. I'm sure you understand that.'

Another roll of the eyes from Charters as Woodes Rogers took up the narration again. 'His Majesty has most graciously extended the hand of charity to these vile creatures, by way of a pardon. He made the offer in September last but set a time limit of one year. Those pirates who do not surrender themselves and accept the pardon by September next will be deemed gallows fodder, and I intend to ensure that their despatch be swift and sure.'

Despite not knowing any of the men involved in piracy, Flynt was still uncomfortable with the fate some might have in store if they didn't vow to discontinue their pursuits. 'But what would you wish me to do?'

Rogers leaned forward to press his point. 'You are bound for Nassau, correct? In pursuit of... what was his name again?'

'Toby Hawke,' supplied Charters.

'Aye, yes, Toby Hawke. I have heard of him.'

'He's a pirate?' Flynt asked.

'No, he is a merchant but he associates with them. Have you heard the names Benjamin Hornigold, Charles Vane, Edward England, Edward Teach? Or Thatch, or Drummond, or whatever name he chooses to use. Better known among his type as Blackbeard?'

'I have heard tell of the latter, but rumour only. Walking the streets of London I have little need to be familiar.'

Rogers laughed. 'Walking the streets of London you rub shoulders with more pirates than you know, sir, and some of them drive fine carriages and sup with nobility.'

The clearing of Charters' throat was more a warning than a means of dislodging phlegm. Rogers waved him away.

'Come, Nat, we know that there are men of business and influence here who profit from theft as much as the scum of the streets and seas. But they do not take the walk to Tyburn and the hempen scarf, but instead frolic with perfumed women and even men, attend the comedies of the Theatre Royal and often appear at court. The damned East India Company has become a law unto itself.'

Flynt was warming to this man, though his comments did bring Lord Moncrieff and the Fellowship to mind.

'They are not the subject of discussion, my friend,' Charters said. 'I feel sure the good serjeant here has pressures on his time, and it would perhaps be a better use of it if you told him what his mission would be.'

'If I choose to accept it,' Flynt said, which prompted another glare from Charters.

'Quite so,' said Rogers. 'These cluster around Nassau like ticks in a field awaiting fresh meat to feed upon. They have colonised the town,

Serjeant, and they sail from its sheltered harbour to prey upon shipping with impunity. The government, being for so long more intent on wars with France, has done little to break their hold, until now.'

'Again,' Flynt said, 'what will you have me do?'

Rogers leaned forward again. 'I need a man I can trust in Nassau.'

'You don't have agents there already?'

'Nobody I trust, and none of them pirates. I need you to go there, seek out information, who is sympathetic to the amnesty, who is not, who would perhaps come to their aid, who is hostile to them. You can relay your intelligences to me when I arrive.'

Flynt turned to Charters. 'And you are agreed? You will not attempt to prevent me from leaving?'

'A threat to trade is a threat to England's wellbeing,' Charters said. 'And these individuals have been allowed free rein for too long. A prosperous nation is a secure one, and we must do all we can to maintain that.'

By England he meant Great Britain, but Flynt didn't waste his breath in correcting him.

Woodes Rogers rose. 'What say you, Serjeant Flynt? Do we have an accord? Will you undertake this mission?'

Flynt considered only briefly. If it meant he would be allowed to depart unhindered, then gleaning what intelligence he could while on the island was a small price to pay. He might even turn it to his advantage.

'In order to elicit information I may have to grease a few wheels,' he said to Charters.

Charters' lips tightened. 'You are going that way anyway, Flynt.'

'Tongues wag more easily when loosened by coin.'

Charters sighed. 'Very well, I shall have funds deposited at your lodgings before you sail. When is that, by the way?'

'I'm currently arranging passage on a vessel headed to the Indies on the morrow.'

Rogers asked, 'With whom will you sail?'

'I hope for a berth on the *Sprite*, captained by a Charles Johnson.'

Rogers' face beamed. 'Charlie Johnson! I know him well... knew him well. He is a good man, an honest seaman. I sailed with him in my privateering days.' He waved a hand towards the scar on his cheek. 'He was there when I took a ball to my face. I made a pretty sum on that expedition, to be sure, but you know I used most of it to pay family debts. Such achievements pale when set against ruin. But this endeavour on which I will embark will not be as ill-starred, I will ensure such. With your help, Serjeant Flynt, I will extirpate this nest of rogues and rid the western seas of their menace.'

—

Woodes Rogers finished his bread and cheese, partook of another glass of wine and then left Charters and Flynt by the rear door.

'Do you believe he can make a difference?' Flynt asked.

'If anyone can, it is he. He is exceeding competent.'

'You've known him long?'

'By reputation, initially, but we have become friendly since he returned from his circumnavigations. I exhort you to do a fine job for him, Flynt. He is the best man for it.'

'He has little regard for the East India Company.'

Charters' lips twitched. 'I thought that would endear him to you. He has been badly treated by them, though he is a businessman himself. When his business failed he set off on another expedition, to Madagascar, where he would purchase slaves to transport to the Dutch East Indies.'

Flynt bristled, his good feelings towards Rogers cooling, but Charters waved a dismissive hand. 'I know, but not everyone has your aversion to the trade, Flynt. It is legal and it is profitable and there is no room for principles in business. But he had another purpose, and that was to survey the island with a view to expunging the pirates who infested it and establishing a colony. The East India Company were none too keen on that notion and so ensured that it was blocked.'

'They were happier with pirates than colonists, I take it?'

'They are most jealous of their influence, and any notion of civilisation might threaten their monopoly in the east.'

Flynt's smile was wry. 'There is no room for principles in business, correct, Colonel?'

Charters shot him a look.

'On that subject,' Flynt said, 'I encountered Lord James Moncrieff last night.'

'I've had no reports of any bodies being found in the streets, so it was not a violent encounter?'

'It was not.'

'Now there's a rarity. I may make note of this phenomenon in my journal.'

'He knew of my stepmother's abduction. He is acquainted with the man Hawke.'

Charters' eyebrows shot up. 'Is he, by God!'

'He had contact with him in Edinburgh at the time and he inadvertently identified Gideon to him.'

'Inadvertently?'

'I believe him. He had no idea of Gideon's connection to him and the true reason for Hawke's inquiry. However, it does suggest to me that perhaps Hawke is connected somehow to the Fellowship.'

'Unless their meeting was to do with legitimate business. Moncrieff has extensive colonial interests.'

'That was not my impression. Moncrieff made a point of seeking me out to tell me of him.'

'And they met in Edinburgh, you say?'

'Aye.'

'When was this? Two, three weeks since?'

'Aye.'

Charters was thoughtful.

'This is meaningful?' Flynt inquired.

'I received notice just this morning that Andrew Wilson was found dead in his apartments at that time.'

It was Flynt's turn to be surprised. 'Murdered?'

'It appeared to be by his own hand. He slit his own throat, the bloody blade still clutched in his hand when he was found.'

'And you believe this?'

Charters affected a shrug. 'There was a rumour that he had been buggering a boy for some time and that the secret was about to break.'

'When did this filter out?'

'After he was found dead.'

Flynt studied the colonel. 'You don't accept it, do you?'

'I have no reason not to accept it.'

'And yet...?'

Charters pursed his lips. 'Knowing now that Moncrieff was in Edinburgh at the time does give me pause.'

'You think he had Wilson assassinated?'

'We know he is Fellowship. We know he has ambition. We know he will not baulk at murder. Perhaps Wilson became an impediment to the former and fell victim to the latter.'

'And what of the boy he was supposed to be buggering?'

'Left the town and the country, apparently.'

'If he ever existed.'

Charters didn't comment. 'There have been two other deaths in the business community. One a board member of the East India Company, the other with the South Sea Company. Both accidental, at least apparently.'

'An epidemic, it seems,' said Flynt, noting the word apparently. Charters didn't believe there was anything accidental, or self-inflicted, about the deaths. 'You will be investigating, I take it?'

'I will have someone look into it. It will not be your concern.' Charters stopped and reconsidered. 'Although perhaps, in some way, it will be. Let us assume that Moncrieff is clearing house a little to pave the way for his ascent to the position of Grand Master, removing anyone who might be more sympathetic to a candidate other than Moncrieff. What if this fellow Hawke was loyal to the late and entirely unlamented Mr Wilson?'

'You think Moncrieff is priming me as a weapon?'

'He knows of your skill as a taker of lives...'

'When I have to...'

'And you may well have to, in this case. He told me once he doesn't like to gamble, but perhaps in this case, he is...'

9

On leaving the Black Lion, he saw Jack Sheppard was no longer alone. He had been joined by Joseph Blake, Jonathan Wild's lieutenant, known as Blueskin for reasons unknown, but Flynt had heard that it was because of the dark hue to his skin, or the constant stubble he sported despite having shaved, or because he once had a partner in crime named Blewitt and the duo was dubbed Blewitt and Blueskin by the witty scamps of the city's criminal class. More likely it was a mix of all three.

He had hoped to escape the tavern without being seen, but this time Jack spotted him and began to raise a hand in greeting. Blueskin caught the movement and glanced over his shoulder at Flynt. He flicked a finger to instruct Jack to lower his hand, then rose to swagger between the tables in Flynt's direction. He shot a brief look towards the door from which Flynt had recently emerged and narrowed his eyes.

'You is often in and out of that there private room, Flynt,' Blueskin said, suspicion dripping from every word. 'I does wonder what it is you gets up to? You got yourself a bobtail up there?'

'I like to eat alone, Joseph,' Flynt lied. 'Mr Hines kindly lets me use his private room when available, to dine.'

Blueskin craned round towards Hines behind the bar. 'Is that a fact? I'll bet he charges you extra for the pleasure.'

'Naturally,' said Flynt, stepping around him. 'Now if you'll excuse me...'

Blueskin manoeuvred himself into his path. 'There has been some coves what has been asking about you. Three of them.'

That caught Flynt's attention and the irritation he felt when Blueskin had blocked his way evaporated. 'Who?'

'Not ones what I has ever had the pleasure previous. They've been at the ask all over the streets, wishing to know about a Jonas Flynt, son of Gideon. I didn't know you had a father, Flynt – I thought you was one of those virgin births, you being so holy and mighty and such. They finally reaches me. They was English but with something else in there what I can't rightly locate. They was tanned like dagoes, too, and one favoured one side, like he was in pain. You caused someone injury lately, apart from me?'

Blueskin still sported the shadow of bruises received during a fight with Flynt. Flynt ignored the questions, turning over what little information Blueskin had provided. Tanned, and one appeared wounded. These men might be the ones who tried to take Cassie in Edinburgh. 'What did they ask?'

'Who you was, what you did, where you frequented.' He paused. 'If you'd been seen in the company of a Black doxy.' He paused again, a sly, almost secret smile giving a curve to his thick lips. 'Of course, I told them about that dusky harlot that does you the tup regular.'

Flynt felt something cold creep over him and he stepped closer to Blueskin, speaking softly. 'What did you tell them?'

Blake didn't flinch. 'I tells them where to find her. They seems to me to be tasty fellows...'

'When was this?'

'Not five minutes since. If I'd known you was above I would have sent them your way. If you run you might catch them.'

Flynt wanted to strike that sly smile from the man's face. He'd wanted to punish Belle for years, since she embarrassed him when he was trying to bully her, and now he thought he had.

'If Belle is harmed, Blueskin, mark me – we shall have a reckoning.'

Blueskin stepped back now, his arms outstretched. 'Why not go for it this here minute, Flynt? I owes you a lump or two.'

He gestured again to the bruises. Flynt had no time, though, to satisfy the man's need for violence. He had to get to Mother Grady's immediately.

'Another time,' he said, brushing past him.

'You knows where to find me,' Blueskin shouted after him, a laugh rippling under the words and young Jack watching the exchange, his face pale, his eyes both troubled and fearful.

From Drury Lane to the corner of Covent Garden, where Mother Grady's house was located, was a short distance and Flynt strode through it, searching for three men walking together. He banged on the door with urgency, darting in when Jerome swung it open.

'Miss Belle, where is she?'

Jerome was flustered by the speed of the question and Flynt's terse delivery. 'She's… Miss Belle is in her room.'

'Does she…' Flynt stopped, swallowed. Damn it, he had to ask. 'Does she have company?'

'She doesn't take gentlemen now, thee knows that, Mr Flynt.'

Flynt stared through the open door towards the Piazza, hoping to see the men staring back at him, but knowing that if they knew their business they would have split up. Three men in a group would be noticeable, three men singly less so. However, nobody took an undue interest in the doorway. They were there, though, he could feel it. 'Has anyone been asking for her?'

Jerome's face creased with concern as he followed Flynt's gaze. 'Nobody. Mr Flynt, what be amiss?'

Flynt ignored the question. 'And Miss Cassie? Where is she?'

'In t' back parlour, with the young gentleman. But, Mr Flynt, I do wish thee would tell me what troubles thee. All is well in this house.'

Jerome knew Flynt too well, and knew that his demeanour was the result of some danger.

'It's nothing, Jerome,' he said, placing a hand quickly on his arm. 'Have there been *any* strangers calling?'

'Nay, business has been quiet all day. Only two gentlemen currently engaged, and they're both regulars.'

Flynt relaxed. Whoever had been asking about him hadn't decided to gain entry to this house. That didn't mean they wouldn't. 'Don't let anyone in unless you know them…'

Jerome, with another anxious scan of the Piazza, closed the door. 'Is there trouble, Mr Flynt?'

Flynt was already crossing the foyer. 'Perhaps.'

He had hoped to avoid Mother Grady but it was in vain, for she caught him as he made his way to the parlour at the rear of the house, where the culls were not allowed. She emerged from the direction of the kitchens, looking more frail than before, but the determined set of her mouth and the flinty glint in her eye assured him that she was as alert as ever and she meant to prove it to him.

'A word, Jonas Flynt, if you please,' she said.

'I regret, Mrs Grady, that I am at the minute most pressed...'

'Whatever it is will wait.'

'I'm sorry, but...'

She seized him by the upper arm and propelled him towards the reception room in which he had met with Cassie the night before. She may have appeared weakened but her grip remained strong, and unless Flynt wished to jerk his arm free and push her from his path, he had no alternative but to allow her to divert him from his course. She had a way of making him forever feel the guilty schoolboy. He shot a final glance towards the closed front door where Jerome leaned, his eye to a small peephole. The man was sturdy and dependable. He wouldn't let anyone gain entry unless he knew them.

Once inside the room, Mother Grady said nothing for a moment but stared down at the cold fireplace, one hand on the high mantle, the other pressed to her chest as if she was feeling for her own heartbeat. He was eager to be on the move and was about to ask her what business she had with him when she slowly turned and faced him.

'I'm dying,' she said, simply. 'I won't see another Christmas. Something malignant grows within my lungs.'

He'd had his suspicion for some time that she was ill, but to hear her declare her mortality so baldly still came as a shock. He considered saying something sympathetic but held back. That was not why she was telling him this, and she wouldn't welcome anything he might say, whether heartfelt or not. Mother Grady seemed to understand his silence and gave him a thankful nod, then coughed. It began softly but swiftly grew harsh and she pressed a handkerchief that she clutched in her hand to her mouth.

They had seldom been in accord, but for some reason he felt immense sadness at the thought of her passing. 'Does Belle know?'

'Not yet, and I will thank you not to inform her.'

He inclined his head to let her know he would remain silent. She nodded, satisfied.

'We've had our differences, Jonas Flynt.'

'We have.'

'But I believe you to be a man of good heart, underneath it all.'

He wasn't sure that was truly the case but he thanked her anyway.

'Who is this woman and child you have brought to the house?' she asked. The question was couched casually but he detected a sharper edge.

'I didn't bring her...'

'She came to see you...'

'Aye, she did...'

'Then who is she?'

He had a suspicion she already knew but wished to hear him explain the situation. 'My stepsister, Cassie, and her son.'

'And who is the father?'

He chose his words carefully. 'His father was an old friend of mine...'

'I don't mean the man he called father. Whose blood is it that runs in his veins?'

When he didn't reply, her head tilted to one side and he saw the flash of her temper in her eyes. 'He's your boy, Jonas Flynt, it's as plain as paint. By God, he even bears your name! And you stand there before me and take no responsibility.'

He didn't wish to go into this, not now. He needed to get them out of this house. 'I don't shirk my responsibility. It's complicated.'

She sighed, the anger gone again. 'Aye, life is complicated, is it not? But yours is often more than that. I said you had a good heart and I believe that, but I also said underneath it all. It's the part on top that worries me. The thieving, the killing.' She waited for him to speak but again he remained silent. She hadn't said anything untrue. 'You don't deny it, that's a good thing. Don't misunderstand me, I know some men are in need of a killing.'

She began to cough, the kerchief called into play once more. The spasm eased, the handkerchief spirited away again. 'I'm not going to lecture you, Jonas Flynt. I've done that too often and I grow weary of it, as I'm sure you are. But you know I remain concerned for Belle.'

'I would never knowingly hurt her.'

'You already have, lad,' she said, sadly. 'I've said before you're not good for that girl, and you must leave her be. Harlot she was but she be harlot no longer. She is a businesswoman now and she doesn't need the likes of you causing *complications*.' She coughed again and this time Flynt heard something harsh hawking up and when she removed the kerchief again he spotted traces of red that was near black on the white linen.

The Admiral had a tendency to cough but as far as Flynt knew there was no blood discharged. Mother Grady turned away for a moment, catching her breath.

'I'm told you will be leaving us to go a-wandering again,' she said, staring into the dark, dead ashes. 'You did such before and I prayed you wouldn't return, but return you did, like a pox that won't let go.' She turned to face him. 'I beseech you, this time stay away. Grant a dying old woman this final wish. Leave Belle be. Leave her to get on with her life, because if you don't...' Further coughing interrupted, more dark blood left on the handkerchief. She swallowed, took a breath, a deep one, then fixed her eyes on him once more. 'If you don't, you will take that life away from her.'

10

His eye fell on the bags immediately he entered the parlour. Though it would save time, he realised that Cassie and young Jonas had been planning to move on, and that annoyed him. His irritation was unwarranted, but all the same he felt it build, though he was aware that he had no right to comment upon it at this point.

'We're leaving,' he said. 'Glad to see you are prepared.' Perhaps he felt some right to comment after all.

'You vanished without a word,' said Cassie. 'I should be used to that by now.'

There was reproof in the harsh tone but she was entitled to an explanation of his absence. 'I've arranged passage. We sail on the morning tide.'

'All of us?' Young Jonas, his face taut, prepared for argument. 'If you intend to leave me behind, then know that I will follow, even if I have to work my passage like Grandfather...'

'We're all going,' Flynt said, earning a reproving look from Cassie, so he had to explain. 'Hawke has men here; they are outside as we speak. I must convey you to a place of safety for the rest of the day. To leave the boy...'

'I am *not* a boy!'

Flynt sighed. 'If we leave Jonas behind then he may be open to some peril.' He nodded to the lad. 'Anyway, I believe he would do as he says and would follow. He's better off in our care.'

'I don't need you to care for me, sir. I can look to myself, and to Mother, quite sufficiently without you.'

Flynt's anger finally revealed itself. 'Damn it, listen to me...'

'I will not listen...'

Flynt advanced on him. To his credit the boy didn't flinch. 'There are men out there who wish to take your mother. They may also take you should they deem you their property, or kill you just to be free of you.'

Young Jonas glared into Flynt's face. 'I understand that.'

'Do you also understand that to evade them you may have to kill? Can you take a life? Can you put a ball into a man's chest? Can you? Do you have it in you to thrust a foot or two of cold steel into flesh? Will you stomach watching the life in his eyes bleed away, knowing that you have taken everything from him, his past, his present, his future, everything?'

'Jonas.'

Cassie's voice, soft, and he had to turn his focus away from the boy in order to see which Jonas she meant. She faced him, imploring. 'Leave him be.'

Young Jonas struggled with his anger, or fear, or both. His face quivered, his eyes awash, his lips trembling. Guilt doused Flynt's own anger but he couldn't let the matter rest. 'Did you bring a weapon?'

Cassie reached into the bag at her feet and produced the pistol Flynt had last seen wielded in the hand of his old friend Rab. 'Can he handle this?'

'I am here before you,' said Jonas, his voice shaking. 'You can address me directly.'

'Well, can you?'

Jonas nodded and Flynt took the weapon from Cassie and held it towards him. 'Then let me see you prime and load that and then we will take our leave.'

Without a word, Jonas delved in the bag for powder and ball. He was about to sit at a small table but Flynt stopped him. 'You won't have table and chair in the street. Work on your feet.'

With a heated glare, the young man set to work, Flynt watching him closely. His fingers trembled but he knew his way around the pistol, if a little slowly.

'You cannot leave without speaking to Miss St Clair,' said Cassie.

He had intended to call upon Belle but Mother Grady's exhortation to leave her had taken root in his mind. He saw the wisdom in her warning that the life he led was not one to inflict on her. Her own life had been difficult enough, though not without its comforts. Until recently, she had been someone else's property and as a youth he had listened as Mercy, Cassie's mother, told him what it felt like for another person, a white man, to hold your life in the balance, that you ate and slept, lived or died at his whim or if it profited him. Mary Grady had granted Belle her freedom but even that was an insult, for it was Flynt's belief – instilled by Mother Mercy – that it was not in anyone's gift to grant another freedom. Then he recalled that on occasion he had attempted to buy her freedom himself, making him even more complicit in the trade.

'We have not the time,' he said, as young Jonas fumbled a little with the ball and wadding, his voice sharp but only as a means to lacerate his own conscience.

'Make the time,' Cassie insisted, her tone as keen as his. 'Don't simply vanish from her life. She doesn't deserve to be treated so unfairly. If you do not then it will prove that you have not changed in any way.'

She would not move from this place until he had done so. Young Jonas completed the loading and priming process and now regarded him with a triumphant but still defiant expression.

'Good,' Flynt said, and the boy beamed. 'But not good enough. It needs to be faster. Out there you will have only seconds to reload, if that.' Young Jonas's expression darkened as Flynt shot Cassie a submissive look. 'Take some time to practise. I won't be long.'

—

Belle was at the small bureau in her room, the quill in her hand scratching at paper. He had knocked before entering, and she gave him the fleetest of glances before returning to her scribbling.

'I sail for the Indies in the morning.'

The scrape of the swan feather point was unusually loud.

'I'm taking Cassie and, em, her son away now to a place of safety.'

Without halting the flow of words on parchment, or even looking in his direction, she said, 'They are safe here.'

'There are men seeking to do them harm. I can't risk your wellbeing or that of the women under this roof.'

'There are always men seeking to do someone harm in your life, isn't that so, Jonas?'

'The world is a dangerous place.'

'Especially when you are around.'

There was a bitter sound to her voice that he should have expected, given her comment the night before, and its coldness stung. 'I must go, Belle.'

'Then go.'

'I didn't want to do so without saying goodbye.'

The quill was dropped and Belle turned to face him in the chair. 'That was unusually thoughtful of you.'

He suspected that she had guessed that Cassie had more or less forced him to do so. Guilt made him close his eyes for a second. 'I'm sorry, Belle.'

'No, you're not, Jonas. There will always be men with violence in their hearts. That's your life, that's what you live for, that's what makes your blood pound.'

'I don't seek danger...'

'No, but you don't shirk from it. It's like opium to you – you say it isn't part of your life yet you cannot live without it. I know this and I could learn to live with it if it weren't for the fact that there will always be Cassie. She is a fine woman. We talked long into the night, and thanks to her I finally understand you. You will never settle, Jonas, you will never be truly happy, not with me. Cassie will always be there between us.'

'That's not true, Belle...'

'I saw it as soon as I entered the room last evening to find you two together. The way you looked at her. I am but a replacement...'

'No...'

'Yes. We could be sisters, she and I. But you couldn't have her, so you settled for me. I will not be second best, Jonas.'

She gave him the opportunity to speak but, mindful of Mother Grady's exhortation, he couldn't find the words.

Belle's face hardened in his silence. She had perhaps hoped he would deny her words but he hadn't. On reflection, Flynt was unsure if he could.

'Go, Jonas,' she said, her voice sorrowful. 'You will return, or you will not. I will be here, or I will not. Go now. Do what you have to in order to protect Cassie and her son.' She fixed him with a steady gaze. 'Your son.'

He felt he had to say more. He wanted to say more. But he couldn't say anything. Instead, he gave her a long look, hoping to convey something of his own confusion, but all he saw was her beautiful face frozen with resolve.

He turned and left the room. As he walked back to the staircase, he heard her close the door.

11

He couldn't pick them out from the people who milled around the Piazza, but he knew the watchers were there, somewhere nearby. It was a sense that had stood him in good stead in the past, a tingle of the flesh that alerted him to danger. As he led Cassie and young Jonas across the square before St Paul's Church, he spotted them. Three men, dressed in thick, long coats much like Flynt's own, converging on them from different points, one limping a little, his hand pressed to his side as if each step caused him pain. He doubted they would attempt anything in such a populated area, for the Garden was filled with tradesmen and their customers, selling and buying fruits, vegetables, flowers, herbs; porters bustling with their loads; carters making deliveries; loungers reclining and loitering; culls in search of pleasure and doxies intent on providing it. However, he couldn't be sure exactly how professional these men were. For his part, he had no desire for action in case any innocent bystanders were hurt, but he had to be prepared.

'Where is the pistol?' he muttered to Jonas.

'I have it,' Cassie responded, looking about her and seeing the men approach.

Flynt nodded with satisfaction, for he knew Cassie to be a proficient shot and had also witnessed how cool she appeared when confronted with danger. The boy was an unknown quantity.

'Is it to hand?' he asked.

'Under my coat,' she said, her eyes coming to rest on one of the men in particular. Flynt saw recognition flicker.

'Do you know him?'

'I believe so. He's older...'

As are we all, Flynt thought, and if events do not go well, we may not grow much older.

The men came to a halt a few feet from them, the one Cassie had recognised nodding to her. He was handsome, his skin tanned, his black beard well trimmed and devoid of any silver hairs. 'Cassie. It's been a long time.'

'Not long enough.'

The glimmer of a smile as he shifted his attention to Flynt, looking him up and down most carefully. 'You'll be Jonas Flynt, I assume?'

His accent carried a twang Flynt had never heard previously.

'I'm Daniel Hawke,' the man said when Flynt neither confirmed nor denied his own name.

So this was Toby Hawke's son. He was of an age with Cassie, in whose direction he raised a finger. 'That is my family's property.'

'I am nobody's property,' Cassie said through gritted teeth.

'The law says different.'

'The law is inhuman.'

Hawke shrugged. 'Maybe so, but it's still the law and Father wants you back. We already have Mercy, as you know.'

'Not for long,' Cassie said.

Hawke laughed. 'You were always a spirited one, Cassie, but I'd curb it when you return home. It doesn't sit too well on the island, property being uppity. It's a habit which we may need cured.'

'Cassie's with me,' Flynt said, and Hawke's gaze moved back in his direction. It was an easy shift of focus, almost lazy. This man was very sure of himself.

'You do speak, I was beginning to wonder if you were mute. We know about you, Jonas Flynt, the man-killer. We first heard your name in Edinburgh, when making inquiries about old Gideon. You made an impression on that town when you were there last. Then just this day we had us a long talk with a friend of yours, fellow by name of Blake.'

'That must have been very dull for you,' said Flynt. 'Joseph is hardly a stimulating conversationalist.'

'He speaks highly of you, too.' Hawke made a show of surveying their surroundings. 'I doubt you will pull those pistols of yours here – too many witnesses. What is it you call your weapons? Tact and Diplomacy, correct? It seems strange to me that a man anoints his weapons with names.'

'I am most eccentric.'

Hawke chuckled. 'That's one word for it. But I don't see you being mad enough to fly lead among all these people.'

'I wouldn't bet your purse on that. I'm an extremely good shot.'

'And we are three to your one, each armed with a brace of weapons apiece.' He leaned forward. 'And just so you know, we are all extremely good shots.'

'The odds are more even than that,' Cassie said, her hand hidden beneath her open coat.

'You have a pistol under there?' Hawke seemed amused. 'Yes, of course you do. My friend here still carries the mark of your aim. Luckily, it hit flesh only.' He squinted as a thought struck him. 'Or do you mean your friend, the one who accompanied you from Edinburgh?'

She hadn't mentioned the man had travelled with them, and Hawke caught Flynt's surprised expression.

'You didn't know she had a protector? He cost us dear in Edinburgh and he was with her all the way to London.' He made a show of looking around again. 'But I don't see him here.'

'He's here,' said Cassie, but the doubt in the way her eyes flitted around them was most apparent. With no small measure of discomfort, Flynt considered again the identity of her mysterious friend, and why she hadn't told him he was in London, but now was not the time to inquire.

Hawke made a show of looking about him. 'And yet he conceals himself. Perhaps he is shy. Or perhaps simply not present.' He ended his search. 'But the hour grows late and our ship awaits at harbour. You know what they say about time and tide. So, if you please, allow me to take back my family's property and we shall be on our way. No more blood need be spilled.' He jerked his head towards young Jonas. 'You can keep the offspring.'

He took a step forward, prompting Cassie to show the pistol though continuing to ensure it remained concealed from the eyes of any onlookers within the folds of her skirt. Her features set firm, her voice steady. 'One more step, and so help me God, I will drop you.'

Hawke halted, irritation creasing his brow. 'I grow tired of this.'

'That is easily remedied. Turn and leave us alone.'

'I can't do that. My father wants you back.'

Young Jonas seemed about to leap forward in defence of his mother but Flynt caught him by the arm and dragged him behind him, then eased Tact from his belt, but kept it at his side. 'Let's all just take a breath. As you say, there are witnesses.'

Hawke seemed amused. 'Very wise, Mr Flynt, but you must remember that we still outgun you. We could have you bleeding to death on these cobbles and be away before anyone knew it. We sail on the evening tide, and it would take the Watch or the Magistrates longer than that to identify us.'

'Ah, friend,' a voice pitched in, 'it won't be Jonas here left bleeding to death, on that you can wager.'

Flynt craned past the three men to find the source of the voice and saw a familiar face grinning at him, two pistols barely concealed in the folds of his black greatcoat, his long blond hair tucked away beneath his wide hat.

The last person Flynt had expected to see was Gabriel Cain.

12

'Hello, Jonas. I see you're making friends again.'

Flynt was torn between his need to maintain his focus on Hawke and his men, and the questions that raged through his mind concerning his old friend. He risked a glance at Cassie but her face was impassive. Young Jonas, though, was smiling; the first time he'd seen the lad in any way welcoming.

They knew each other.

This was the friend who had helped Cassie.

Flynt couldn't help but have questions. However, they would have to wait until this current situation was dealt with.

'As you can see, the odds are not quite in your favour as you believed.'

He forced a light quality into his voice, even though he was both confused and suspicious and, yes, jealous.

Gabriel Cain. Former highwayman, possible killer for hire, unapologetic womaniser.

Dan Hawke had turned to glare at Cain. In return, Cain gave him a broad smile and even a wink. Hawke's two companions were unsettled by this turn of events, for they shot one another an anxious look while their feet suddenly became restive.

Flynt addressed his remarks to them, now adopting a reasonable tone. 'You don't want matters to become unpleasant, gentlemen. Mr Hawke here has a personal interest in proceedings but I hazard that you are merely on wages.' He discerned from their expressions that he had hit the mark. 'Believe me when I say that Mr Cain and I are most experienced in activities such as this, and we always emerge upright and breathing while others have not. I urge you to consider your own futures. One of you, perhaps both, will not live to fight another day. Back away now, while you can.'

When the men displayed uncertainty, Cain added, 'Pay heed to my friend, boys. This offer will not be repeated.'

'Stand your ground,' Hawke ordered, but it was evident even he was considering his position. He had Flynt and Cassie ahead of him and Cain to his rear. If hostilities erupted, he would be caught in a crossfire and that was never an ideal situation. However, he was determined not to give in gracefully. 'You are harbouring stolen property, Flynt.'

'That is a point we can debate until the end of the earth but one on which we will never agree. Only you can decide whether it's worth dying over.'

Hawke stared first at him, then at the still-grinning Cain, then at his men, who were already edging away. He grimaced, then spat onto the cobbles and sneered. 'I could take it to the law.'

'Be my guest, but you would miss the evening tide,' Flynt pointed out. *In any case*, he thought, *we will be gone from England by this time tomorrow.*

Hawke delayed as long as his pride would allow before he also took a few grudging steps back. As Flynt motioned for Cassie and young Jonas to move ahead, the man's eyes didn't leave Cassie. Before she moved on, Cassie gave him the challenging stare Flynt knew so well. He backed away in her wake, not letting the men out of his sight, always prepared to raise the pistol in his hand. Hawke was already berating his accomplices, almost certainly bolstering their courage for pursuit, and without doubt informing them that they would not be paid unless they did so.

'Swiftly now,' Flynt said over his shoulder, 'move!'

They didn't run, for such action would only raise suspicion among the people who milled around them, but their pace wasn't far from it. As they veered into Henrietta Street, Flynt looked back to see that Hawke's exhortations had produced the desired effect, for he and his two men were in pursuit. Their pace was brisk, so Flynt and his party increased theirs to reach the end of the street, where they turned left and crossed to the opening of Dawson's Alley that linked with St Martin's Lane to the west. It was only about two hundred yards long but it was narrow and more easily defensible, even though they could not stand shoulder to shoulder. Flynt brought them to a halt halfway, Cain at the rear keeping watch, but there was no sign of their pursuers.

'They weren't far behind us,' Cain said, 'so where are they?'

Flynt had thought they would blindly follow, but that wasn't the case. 'If they are clever they are making a plan. They will know that the first man to show himself will be the first man to fall.'

Cain smiled. 'Perhaps they are drawing lots.'

'Or they have split up,' Cassie ventured, 'one party taking another route to approach from the other end of the alley.'

Flynt and Cain exchanged glances, each knowing that if that were the case, then they would be caught in a crossfire. 'Yes, that's what I would do,' Cain said.

Cassie glared at Flynt. 'Then you have boxed us in worse than before.'

Entering the alley was a chance he'd felt he had to take. He had gambled that it would be empty; it being cramped and even in daylight ill-lit, it was a thieves' paradise, so there was little danger of an innocent being injured.

Flynt ignored Cassie's reproving look to squeeze past her, his back brushing against the high brick wall, his attention never wavering from the narrow opening. 'If I fall, be ready to use that pistol.'

'I'll use it,' she said, her mouth a tight line, 'and I hope to God it's Dan Hawke who steps across its barrel.'

'What about me?' Young Jonas's voice was strong, but tight.

Cain handed him one of his two pistols. 'Take this. Remember what I taught you. Aim for the centre of the body, don't attempt anything like a head shot or to simply wound for you will surely miss. This is no time for scruples, lad.'

The boy accepted the pistol with a warm smile and positioned his feet to make steady his stance. So Cain had been teaching him how to shoot. He really was a friend to the boy and Cassie. Again Flynt felt a surge of suspicion and jealousy, which he had to force down. He had no right to feel either. All the same, he gave Cain a strong look as he watched him reach to his back under his coat and produce a third pistol. That, at least, made him smile a little.

'When did you take to sporting three weapons, Gabriel?'

'It never hurts to be prepared for any eventuality.'

'You always said if you can't manage with two shots, then you are doomed.'

Cain gave him his smile. 'A man grows older, Jonas.' He faced the alley mouth once more. 'As you will without doubt already know.'

Despite his reservations about Cain growing so close to Cassie and her son, Flynt was glad to have his old friend back. They had been through a lot together, and even though he also harboured misgivings over how Cain had made a living since they had first parted ways, he knew that together they could handle almost anything, older or not. He drew both pistols and readied himself, hoping that he had not indeed boxed them into this confined space.

They waited in silence. Footsteps echoed from the streets along the narrow passageway. Figures passed at each end of the alley, but none entered, nor even glanced into the shadows. The sound of hawkers crying their wares and cartwheels rubbing over cobblestones reached their ears. The hooves of horses clipped. Above them, the sliver of sky beyond the rooftops grew dim. He risked a glance behind him, caught a look passing between Cassie and Cain. Was that an affectionate light in her eye? Her gaze was most certainly softer than she had given him since she arrived.

Stop it, Flynt. You gave up any right many a long year ago.

He concentrated on the end of the alley.

'Could they have given up?' Young Jonas's voice seemed hopeful.

'They're waiting for us to show our faces again,' said Cain.

'So what is our next step?' Cassie asked.

Flynt craned over his shoulder again and caught Cain's eye. 'What do you think?'

Cain considered. 'There were three of them...'

'That we know of,' Flynt added.

Cain accepted that. 'We must proceed with the assumption that there were no more. So if they did separate, it follows that only one waits at one end and two at the other. Either way, it makes our odds so much better.' He grinned. He lived for such moments. 'I think we show our faces.'

Flynt nodded in agreement and began to lead them in the St Martin's Lane direction, both pistols held before him should Hawke or his men appear. It was only a few yards but it seemed like a mile. He knew that he presented an open target in the confined space, as Cain did at the rear, so if there was any sign of attack there could be no hesitation. As soon as an aggressor revealed himself, he would shoot and shoot to kill. As Cain had told young Jonas, this was no time for scruples.

The end of the passage presented itself at last and he paused. None of the people about their business paid attention to the tall figure standing in the shadows of the lane, but if he stepped out with weapons aloft it would cause some consternation. He hadn't thought this far ahead, it was true, but they were here now and he had to proceed. He took off his hat and handed it to Cassie behind him. She was about to speak but he shook his head, then pressed his back against the wall to his right and tilted his head as far as he could without it protruding beyond the brickwork. From that vantage point he could see a little further down St Martin's Lane towards the Strand. Nobody lurked within his field of vision. He then did the same in the other direction. Still no sign of a threat, but that didn't mean they weren't pressed hard against the wall and lying in wait.

He took a deep breath. There was nothing for it, he would have to step out, but not with Tact and Diplomacy in evidence. He would have to lower them and trust his speed to win the day or, should he fall, that Cassie or Cain would succeed where he failed. He concealed the hands holding the pistols in the hanging folds of his coat, and breathed in again through his nose. In and out, steady and even, until the flutter in his gut stilled. If this was to be done, it was best done quickly.

He stepped into the wider thoroughfare, every nerve tensed and ready to bring the pistols up.

Blueskin Blake leaned with one foot propped against the wall, one arm across his chest, the other cupping the bowl of a pipe, the stem between his lips. He looked Flynt up and down, amusement dancing in his eyes. 'I was beginning to believe you had taken up residence in that there alley, Flynt. Or had found yourself a doxy for a tuppenny upright.' His eyebrow rose as Cassie emerged. 'I was in jest, but now I sees I has been correct.'

Flynt ignored the remark. 'What the hell are you doing here, Blueskin?'

'I reckons you had some trouble and I thought to give you my assistance.'

He jerked his head further up the lane, where a decidedly discommoded Hawke and his men stood under guard of four of Blueskin's men, including Jack Sheppard.

Flynt was puzzled, his relationship with this man not being one of mutual support. 'Why would you do that?'

Blueskin pushed himself free of the wall and rapped his pipe on the brickwork to empty it. He then carefully placed it in his pocket. 'It's my public duty, ain't it, as one of the Thieftaker General's men, to nip any wrongdoing in the bud.'

Flynt's nerves escaped with his laugh. 'Of course it is.'

Young Jonas and Cain were now also in view and Blueskin studied them carefully.

'Tell me the truth, Joseph, why did you do it?'

Blueskin's attention returned to him. 'Because I don't like being beholden to you, Jonas Flynt. You spared my life that night under Whitechapel Hill.'

'That evened the score between us, and you know it. You had already saved me from the Thames.'

'I don't like the score being even. I much prefer it being you that is beholden to me.' He waved his hand towards Hawke. 'And now it is the way I likes it.'

Flynt almost laughed again. No matter, he had saved them and for that he was grateful, even though the sight of young Jack forming part of his crew remained disturbing.

'Thank you, Joseph.'

Blueskin grimaced. 'I didn't do it for your thanks, Jonas Flynt. As I said, I had reasons of my own.' He jerked his head towards Hawke again. 'What shall I do with them?'

'Don't harm them.'

'I wasn't intending such. They ain't done me no hurt.'

'I would be further in your debt if you held them until we are well away.'

'How long?'

Flynt had no idea when the evening tide was but he wanted Hawke to miss it, if only as punishment. 'Until tomorrow afternoon? Can you do that?'

Blueskin sneered. 'I can keep them as long as I likes.'

'Till afternoon will suffice. I am grateful, Joseph.'

'Keep your grateful. You is a rum one, Jonas Flynt, I says it often. And I don't like you, not one bit. But I'll say this, you is most capable, and there may come a day when I needs a most capable man.'

'As long as you're working for Jonathan Wild, that day may come sooner than you think.'

'That ain't for you to judge. Mr Wild treats me fair so I treats him square and you ain't got no call to cast no absertions. I won't hear no detrimentals to the contrary.'

Flynt was fairly certain he meant aspersions but he didn't correct him. 'I have one more favour to ask, and it would place me even deeper in your debt.'

Blueskin's eyes lit up. 'That I do likes the sound of.'

'Young Jack,' said Flynt. 'I need you to look after him for me. Thief he may be but he's not like you or I. He is street canny yet still very much an innocent. Our world is dangerous, I don't need to tell you that, and he needs someone to watch over him.'

'I will do such, but not as a favour to you. I likes the lad, too. He's bright and is most nimble with the dipping. He has a skill with locks, too, which is exceeding useful. You has my oath that I will watch out for him.'

Flynt accepted the pledge with a nod and motioned to his party that they should head towards the Strand. As he turned he heard Hawke's voice calling after him.

'Jonas Flynt, this doesn't end here. My father will have his property returned, either upright or in a box.'

A low noise grated in Cassie's throat as she whirled to dodge around Flynt and stride towards Hawke. Flynt motioned for Cain to stay with young Jonas and followed her. He didn't believe Hawke posed a danger, not surrounded by Blueskin and his men. He was more concerned with what Cassie might do.

'What happened to you, Dan?' she asked.

He seemed taken aback by the question. 'What do you mean, what happened to me?'

'You didn't used to be like this. You were different from the other white boys. You were kind, sweet. When did you change?'

Hawke's eyes hardened. 'When Gideon Flynt stole my father's property and left him for dead. I learned my place in the world after that. I learned your place in the world.'

Cassie was silent for what seemed like a long time. Finally, she took a further step closer and Flynt readied himself to grab hold of her if she lunged. The light was dying but St Martin's Lane was still thronged with people and it wouldn't do to have an altercation in public. As Charters had pointed out, slavery was legal and, distasteful though it was, Cassie was indeed the property of the Hawkes. If she assaulted him before witnesses and the constables were called, it would not go well for her. The law, and society in general, did not share his aversion to the trade.

Cassie said nothing further, nor did she make any move to strike Hawke. She simply stared into his face for a few moments, her eyes conveying contempt, and though he attempted to match the intensity of her gaze, he lost focus and looked away.

Hawke made an attempt at saving face. 'Next time we meet, Cassie, you will not be so lucky.'

She turned away without a word and rejoined her son and Cain. Flynt faced Hawke again. 'You had best pray we don't meet again, Hawke. Because if we do, I will not be so restrained.'

Hawke found his sneer. 'You're nothing but a jumped-up footpad. Do you believe me to be fearful of you?'

Flynt thrust his face closer to Hawke's. 'Come at Cassie again, and you will know what fear is. On that you have my pledge.'

He didn't give the man the chance for further badinage. Instead he turned and followed Cassie.

13

'You were fortunate that all worked out well,' Cassie said as he led them towards the Strand.

It was true; he had been reacting to stimulus rather than formulating any kind of plan and it was only by dint of luck that they had emerged from Dawson's Alley without any serious incident. Flynt had occasion before to bless his gambler's luck, and this time it had appeared in the unlikely personage of Joseph Blake.

With Hawke and his men safely in Blake's tender care, Flynt was now free to guide Cassie, Jonas and Cain to his lodgings in Charing Cross. They would spend the night in the coaching inn, then take ship in the morning. Nevertheless, he told them to linger a distance from the entrance to the Golden Cross while he made a sweep of the street and the stables at the rear, just in case Hawke had despatched men separately.

When they entered, Mrs Wilkes signalled to him that she needed to speak with him.

'There was a gent here earlier, delivering something for you,' she said, her voice just a little above a whisper. 'I stowed it in our private apartments, because I didn't want it to be sitting out on display, or even leave it in your room.'

Leaving the others in the bar, she led him through the kitchen area to the rooms she shared with her husband, where she fished a key from the pocket of the smock she wore and unlocked a weathered trunk. She produced two bags bulging with coin. Charters had delivered as promised.

Taking them from her, Flynt said, 'I'll be leaving England tomorrow, Mrs Wilkes. I have business in the Indies.'

'Terrible hot and sweat-filled there, I hear,' she said as she locked the trunk again. 'Will you be wanting me to keep your room for you?'

Flynt had considered more than once moving on from these lodgings. The problem was he was comfortable there, even though too many people knew where he laid his head. Sooner or later that would prove deadly, if not for him, then perhaps for Mrs Wilkes and her husband.

'I would understand if you'd wish to open it to another customer,' he said.

'We doesn't mind keeping it for you. You'll need somewhere welcoming to lay your head when you comes back.'

He reached into one of the bags for coin and she backed away, her hands up as if warding off evil. 'We doesn't take no payment from you, Mr Flynt, you knows that right well. You has helped us out on more than one occasion, and we is happy for you to use that room as your residence for as long as you pleases.'

He had seen off rogues who demanded payment against the possibility of their premises being set to flame. He had also acted as an unofficial bully when patrons in the tavern turned unruly.

'I thank you, Mrs Wilkes, but please...' he dropped a handful of silver on a table beside him, '...take this as a holding fee and for cleaning duties. I would feel more comfortable knowing that you are making some form of income from the space in my absence.'

For a moment he thought she would continue to refuse but then she nodded. 'Then we is agreed. We'll keep your room for you for as long as it takes.'

He returned to the bar and led the others through the myriad of corridors and stairways that made up the Golden Cross, to his room. Cassie stood in the doorway for a moment before entering, staring at the neat little bed, a dresser with water bowl and basin, a solitary wooden chair and a heavy trunk containing Flynt's clothing – two more white linen shirts, black breeches, undergarments, hose as well as powder and ball.

'This is where you live?' she asked.

'This is where I sleep,' he said.

'On occasion,' Cain added.

Cassie walked around the cramped space, before sitting on the bed, pulling an appreciative face at the cleanliness of the bedding.

'It is comfortable, if somewhat spartan.'

'I have few needs.' He jerked his head to Cain. 'Gabriel, let's have a turn around the street to ensure nobody lurks that shouldn't be lurking.'

Cain understood but looked at Cassie and the boy.

'They're safe here,' Flynt said. 'Unless they know which room, I doubt anyone would find their way here. As you now see, the Golden Cross is a labyrinth. And we won't be far away. In any case, I do believe that Hawke's guns have been spiked.'

'For the moment,' Cassie said.

Flynt agreed but didn't say so. He led Cain from the room and down the hallway, then via a dark back stair concealed behind a thick blanket to reach the ground floor and through the inn's front doors onto the street. Only then did he speak again.

'So,' he said.

Cain knew what was expected of him. 'You wish to know how I came to meet Cassie and her son.'

'It has crossed my mind, yes.'

'Put simply, I sought her out because I was curious.'

'In what way?'

'About this jewel of womanhood who had so captivated my friend. So when we parted ways at Gallowmire, I headed north in search of further prospects, life in England having grown tedious for me. I had expected you to do so, too.'

Two years before Flynt had sat at a literal crossroads outside the little village in the north of England and pondered which way to go: to Scotland and Cassie, or south to London and Belle. He chose the latter. Perhaps it was the wrong decision, though he still doubted Cassie would have welcomed him back. Too much water had flowed under that bridge, too much blood had stained that water.

'How did you meet her?'

Cain stared across the street to the statue of Charles the First astride his horse as if he was preparing to gallop down Whitehall to avenge his own death. 'In your father's tavern. You had spoken of it, of him, and I used it as my hostelry while I sought opportunities.'

'Opportunities for the Wraith?'

He had heard of a man known only as the Wraith, who killed for money, but who kept his true self well hidden. During a mission for Colonel Charters, he had pondered whether Cain might well be he.

Cain dismissed that with a wave of his hand. 'I told you before, that individual doesn't exist. He is but a legend, an amalgam of various men and women who walk the killer's path.'

Flynt didn't know whether to believe his old friend, but further pursuit was stifled by Cain's broad smile. 'And while we talk of soubriquets, what of the Paladin?'

Flynt almost groaned. The title had been applied to him by a mischievous woman who was both friend and foe. 'You have heard of this?'

'I had barely set my arse down in a tavern before I heard of this paragon of virtue, this defender of the poor and the oppressed. My God, he makes Robin Hood seem like a robber baron!'

'Like the Wraith, he doesn't exist.'

Cain gave him an exaggeratedly reproving look. 'Jonas, Jonas, Jonas. If ever a man was suited to be deemed a knight errant, it's you.'

'I have committed too many sins.'

'As have we all, but I believe you seek redemption.'

'No. There is none, not for such as you and I. The cognomen was a means to both embarrass and perhaps neutralise me in my work for the Company of Rogues. It's meaningless, and will die out while I am abroad.' Flynt wished to steer the conversation back to Cain's time in Edinburgh. 'And so it was at Gideon's you met Cassie?'

'I knew it was but a matter of time before she would appear there. I recalled you telling me of the shoemaker's workshop on that curiously shaped street...'

'The West Bow.'

'Aye, that be the one, but I didn't deem it decent to visit there.'

'That's uncommon proper of you.'

Cain grinned, for his reputation with the ladies was not without its more devious moments during which the words decent and proper could not be applied. 'I have my moments of decorum. Not many, I grant you, but a few. This was one of them. Anyway, at that point my interest was not in any way connected to any kind of sexual conquest, but mere curiosity, as I said.'

Something quickened Flynt's blood. 'At that point?'

Cain smiled. 'As soon as I saw her I knew why you were so besotted. I admit, I felt the same way.'

Silence fell between them as Flynt took this in. He didn't understand the emotions he felt. Jealousy. Rage. Sadness. Understanding. All mixed together into one broiling maelstrom.

'Did you...?' he began. 'I mean, you and Cassie...? Was there...?'

Cain let him flounder for a time, then laughed. 'Did Cassie and I tup?'

Something stabbed at Flynt's innards and he nodded.

Cain's face was the picture of innocence as he allowed the question to dangle for a few moments, then he took a deep breath. 'We are both men of the world, are we not? And there are moments between such men and women that cannot be denied...'

Flynt winced and something burned behind his eyes. He knew he had no right but he couldn't help it.

Cain saw this and laughed again as he clapped him on the back. 'I was a perfect gentleman, Jonas, on that you have my solemn oath. I made no importunate moves upon her. You may rest assured that I was stricken with an attack of decency.'

Relief doused the burning and eased the tingling of his flesh. He forced a smile. 'That must have been a novel experience for you.'

Cain was thoughtful. 'Do you know, it was, and for a time I thought I had been overcome by some kind of malady. I confess, it's not something I would wish to repeat, for – and I'm sure you'll forgive me for saying this – she is a most beautiful woman.'

'She is,' said Flynt, almost sadly.

'But,' Cain continued, 'you are my friend, and I told you once before that there are damnable few men who I call that. I became a friend to Cassie.'

'And to the boy.'

'Aye, and to the boy. He needs his father, Jonas.'

Cain's words were delivered with heavy emphasis.

'His father is dead and gone,' Flynt said.

'We both know that isn't true.'

Flynt swallowed. 'The father he needs is gone.' It was not a subject he wished to be explored further. 'Did you tell her that you were my friend?'

'My moment of decency extended even to candour in that regard. I felt I could not, should not, lie to her about anything.'

'What about your way with the ladies, did she witness that?'

'She did, and she accepted that a man has his needs. As do women.'

'Especially when their husbands are absent.'

'God has blessed me with many such opportunities and it would be an affront to Him if I were to ignore them.'

Flynt couldn't help but laugh. 'I think somehow that it is a blessing from an entirely different source situated considerably south of heaven.'

'What would you know, for you are a heathen, Jonas Flynt.'

He couldn't argue with that. 'Where were you when Hawke's men took Mercy and came for Cassie?'

Cain exhaled heavily. 'I was, regretfully, sharing the bed of another lady of the town, one whose husband was conveniently away on business. I heard about it the next day but by that time Gideon was in pursuit.'

'And you assisted Cassie and Jonas later?'

'We knew they would return, so for the next two nights we took it in turns to sleep while someone kept watch. They came back but we saw them off again. I do think they were most perplexed to find me facing them down.'

'And what then?'

'The letter from Gideon arrived, Cassie and Jonas caught the first coach to London and I followed, keeping a weather eye open for any more rogues. There were none.'

'None that you saw.'

Cain's look was reproving. 'Come, Jonas, you have seen the quality of Hawke's help, recruited from the dregs of the islands, I'd say, or at least the back alleys of London and Edinburgh. He would have made inquiry and learned of her departure, just as they learned of you. You made quite an impact when you were last there.'

He thought of a bloody riot at an execution in Edinburgh's Grassmarket and the liberating of a friend from the Tolbooth, the town jail, that went tragically wrong. His name would have been bandied about the town and, coupled with Gideon's own high standing within the community, Hawke would have easily learned that he had returned to London and made an accurate assumption as to Cassie's destination.

'They were of sufficient competence to take Mercy,' Flynt pointed out.

'They had surprise on their side – Gideon was absent and Hawke the elder himself was involved, which perhaps made the difference.'

'What do you know of him?'

'Nothing. Cassie has been most circumspect concerning the man and his family. Perhaps she remembers very little.'

'She was eleven years of age when Gideon brought them home.' Flynt thought of her exchanges with Dan Hawke. 'She remembers. You taught Jonas how to shoot.'

'Aye.'

'And Cassie allowed this?'

Cain's face crinkled. 'Allowed ain't perhaps the word, Jonas. The boy and I agreed that his tutoring should be performed in secret.' He gave Flynt a pointed look. 'She has other plans for him, and she fears that there lies within him a trace of his father.'

Flynt hoped that was not the case. 'She wishes him to be lawyer or physician.'

'She does, and he has a fine mind in that head. He will excel in either profession.'

Pride surged in Flynt's own mind. 'What did she say when she discovered you had been training him?'

'It's safe to say that her ire would have been a wonder to perceive had I not been on the receiving end. Like a storm, it's something best to watch from a high hill at a considerable distance. She has no wish for young Jonas to have anything to do with weapons.'

'But you won her round, I'll wager.'

Cain's charm was something that could placate even the most fiery of women. 'Eventually. I pointed out to her that there will come a day when the world is totally civilised and weapons are not necessary but until that day comes, a man needs to at least have a working knowledge of their use. I told her there are three types of men: those who give, those who take and those who explain to those who take the futility of their ways.' Cain looked back towards the door to the inn. 'And it is as well that I did teach the boy, for something tells me we will need all the firepower we can muster for what lies ahead.'

Part Two

At sea

14

Captain Johnson was not a young man. Flynt estimated his age at perhaps fifty, though he could be older or maybe even younger. His shoulders were broad, his forearms bulged with evidence of many a rope pulled and many a sheet hauled. The grip of the sword at his side had been wrapped in leather cord, and was worn near white. It was a weapon that had been used and, given Flynt could discern no wounds on his face, it had perhaps been used well. Of course, he could not tell whether his body was scarred, for he wore a thick blue coat of wool and his head, devoid of wig and from what Flynt could see also of hair, bore a battered three-cornered black hat, even indoors. The skin of that unblemished face, save for the shadow of beard, carried a deep burnish that spoke of many years not just under foreign suns, but also of exposure to the rigours of far less hospitable conditions around the world. He bore an aura of command that came from dealing with hard men all his life. Woodes Rogers had told Flynt that he had sailed with him in his privateering days, and now he captained a merchant vessel, albeit one that was armed sufficiently well to wage a small war. Flynt counted eight guns, all six-pounders, as well as a nine-pound stern chaser. Plying the middle passage westward was a dangerous business, thanks to pirates, and a commercial vessel had to have the ability to defend itself, and so those who manned its decks needed to be capable men. To that end, the crew of the *Sprite* made the most scurrilous of London street villains appear effete. They sported weather-beaten faces, some scarred, but that wasn't unusual on land or sea, and their accents ranged from all corners of the British islands as well as a few from sunnier climes.

Flynt and his party dined with the captain in a small room beside his cabin, the lamp above the table swinging easily with the movement of the ship, sending their shadows tilting and shifting against the wooden interior. The timbers around them creaked and groaned and above them the ship's bell clanged to announce a change of watch. The meal of beef – belly timber, Flynt had learned they called it – and vegetables with doughboys – hard flour dumplings boiled in seawater – was basic but filling. They had been served by a lad who Flynt knew only as Barbecue, who had then cleared away the plates and left them to their port wine. Captain Johnson had kept his head covered throughout the meal and only when the boy had left did he take off his hat, revealing the result of an old sword slash running from just above his forehead clear to the back of his head.

Captain Johnson rubbed his hand over the scar, knowing that they couldn't fail to notice it, but did not refer to it. Instead he fixed his blue eyes firmly upon Flynt. There was a sternness in that gaze, but it also revealed that the man possessed an insight that would detect even the slimmest of prevarications. This was a man who had spent his life trapped on floating wood in all manners of seas with men of varying degrees of honesty and virtue, so he would know instantly should there be any kind of duplicity.

The captain cleared his throat. 'So, Mr Flynt, what are you about?'

He spoke clearly, his voice as rough as the seas they had thankfully left behind, but there was a cultured quality to it that had surprised Flynt when he first heard it on the quayside a week before at the Pool of London before they boarded the pinnace that had carried them out to the *Sprite*.

'What do you mean, Captain?' Flynt asked, knowing well what the man meant.

Captain Johnson drained his glass of port and poured himself another, first offering the bottle to Cassie, who declined, then to Flynt, who laid his hand over the rim of his glass, then to Cain and young Jonas, who both accepted. Johnson replaced the bottle on the table before he spoke again. 'I mean, sir, why do you and your party venture to New Providence Island? It's no place for decent folk.'

Cain grinned. 'What makes you believe we are decent?'

Johnson returned the grin. 'I accept that we are not overly well acquainted. But my employer has vouched for Mr Flynt here, and as you, the lady and the young gentleman sail in his wake, as it were, I believe that such acceptance extends to you.'

Flynt played with his empty glass. 'Captain, you do know of your employer's reputation, do you not?'

'I know him to be a good employer, and that is all I need to know. Though I will say this: his reputation be damned, for there is more honour in that man than in many a godly person. He instructed me to give you and your party passage, though I confess I was surprised not only to see it being four in number, when I was forewarned of three, but also one being a woman, begging your pardon, madam.'

Cassie smiled her pardon. 'You are not in the habit of carrying female passengers, Captain?'

'There have been a few, I'll grant you, but I will speak plain, none of your heritage. And that, sir, is why I ask you again as to your destination, and would have inquired previously had we not encountered that squall on leaving English waters.'

'Squall?' Cain blurted. 'That was no squall, Captain, for it damn near had me wearing my innards on the outside.'

They found themselves in the teeth of the storm as soon as the *Sprite* cleared Land's End, the most western tip of England. The sturdy square-rigged brigantine was tossed hither and yon by the rough seas of the Atlantic and though Flynt had thought his gut capable of handling such rigours, he was swiftly proved wrong. His companions, and at least two mariners, all suffered a similar malaise, but as he retched and heaved, that was of little comfort.

Johnson's lips twitched at Cain's description. 'You're a landsman, sir, so you have little conception of how mercurial the sea can be in its moods. That was a squall, and a minor one at that, so you should pray to whatever God you follow that you encounter nothing stronger. You don't have the stomach for it, sir.' He held up his hand. 'I mean no disrespect. It takes a certain type of man to have his comfort aboard ship, and it does you no disservice if you cannot grow accustomed to it. If we hit real weather, and I pray to the Almighty that we don't, you will know all about it, by God you will.'

Young Jonas inquired, 'Have you ever encountered anything stronger, Captain?'

A slight smile crept into Captain Johnson's eyes. 'I well recall my first encounter with nature's might, yes, as if it were yesterday. In the South Atlantic, it was, and the wind came up on us of a sudden, a hurricane as sure as I'm seated here, and by God, don't you know that it struck us full broadside, and us with our sails still unfurled, for our master was too much infatuated with wine to order them to be reefed.

Pitched right over, we did, until keel was topsides and topsides was keel and we were hanging onto the rails and masts for our very lives. We might've drowned, should have, if truth be told, but another gust righted us again and not a man was lost.' He leaned forward, that smile still dancing. 'But I'll tell you this, young sir, as I hung there, my chest bursting as I struggled to retain the breath in my lungs, I saw something staring back at me from the darkness.'

Jonas leaned forward. 'What did you see?'

'The sea belongs to old Davy Jones and it was him, large as life. His eyes burned as he swam out of the depths and looked us over, but he saw me looking at him and he came right up to me, near as I am to you now, lad, and he stared at me, long and hard, and I tell you, I felt those eyes a-searing right through me.'

'What happened?' young Jonas asked, his own eyes narrowing.

'I'll tell you, and this here's the Lord's own truth, but he wagged a finger at me thus...' Johnson raised a hand and moved his forefinger to and fro '...to tell me that he wasn't going to take me nor my shipmates, not this time.' He looked towards a small porthole. 'But he's out there, old Davy, a-waiting for me, and one day, he'll have me...'

There was silence in that small room then, with both Flynt and Cain holding in laughter, Cassie hiding a smile, as the boy considered the tale. 'Interesting, Captain, but my grandfather told me the exact same tale when I was a child. Davy Jones seems to be most merciful for a sea devil.'

Johnson laughed and slapped the top of the table with his hand. 'By God, who would have thought that the same thing could have happened to your grandsire? It defies belief, it does.' His attention flicked back to Flynt, the humour still dancing from lips to eyes. 'You, sir, recovered more quickly than Mr Cain here. I saw you about the deck, and believe you could be one landsman who might take to life at sea, should you choose to do so.'

The ship's pitch and roll, coupled with the stench, had forced Flynt to seek some respite topsides, where the air was replete with a salt spray that stung at his face and eyes. Lifelines had been run the length and width of the decks to give the sailors something to cling onto, for the waves were like fingers reaching out over the gunwales to pluck the unwary and carry them off. The water surged across the deck to cascade through the scuppers, and the force of it could rip the footing from under even the most experienced mariner. The air was certainly fresher but the yaw of the ship as it bucked the waves remained discomfiting, and the low clouds and overall murk prevented him from finding the horizon, for he had been told by one experienced hand to fix his gaze on it as a means of combatting the *mal de mer*. He returned below, where Barbecue, unaffected by the turbulence, was dishing out quantities of wormwood and wine to combat the nausea.

Flynt inclined his head in gratitude. 'I thank you, Captain, but I believe I will remain on dry land as much as possible.'

'Well, as I say, it's not for everyone. But now, and without further preamble, will you tell me why you wish to reach New Providence?'

Flynt again dodged the question. 'How much do you know of the island, Captain?'

The sense of fun vanished in an instant. 'All I need to know, which is more than most. The waters around Nassau town make a good berth, if you can stomach the company you will keep.'

'The pirates.'

'Aye, that be the truth of it. Pirates they are. I've made a particular study of the Flying Gang, as some of those rogues like to call themselves. Scum, is what I call them.'

Young Jonas now seemed impressed. 'You've had experience of them?'

'Not as much as some, for here I am, lad, still hale and hearty.'

'My grandfather told me of pirates. He'd experience of them when he was at sea.'

He certainly did, Flynt thought.

'He said they were men free of the encumbrances of civilisation,' young Jonas continued. 'He said they were not all killers.'

'Some are, some aren't, but all are thieves and rogues, is what they are, lad, and not men that should be lionised. Criminals, they are, and by their own choice, mostly, and as such they will see only one end and that isn't a pretty one. Nassau town is infested by them, so much so that any decent folk on the island have moved inland to avoid their stench. All governmental authority has withdrawn, and they treat the place like a damned pirate kingdom, and England doesn't do a damn thing to stop them. Pardon my language, madam.' Cassie smiled and nodded. 'They have named their fleet the Flying Gang for they sweep from that sheltered harbour to raid and loot, and with impunity.'

Not for long, Flynt thought, if Woodes Rogers had anything to do with it.

The captain made a point of studying Flynt and Cain closely. 'I've watched you, both of you, and I'd hazard that you are not travellers, nor less colonists, nor planters. You both carry an air about you, something I see in only a few men, and given your connection with my employer I suggest that you are not men who are unacquainted with violence.'

Cain replied with a tiny inclination of his head as he sipped from his glass.

'And so I ask you for a third time, sir, what are you about?' Johnson pressed. 'And please be so good as to oblige me with your reply and not further diversion, for I plainly see that you are unwilling to impart such intelligence. However, I must insist on hearing your purpose. I do not ask out of mere curiosity, though I confess it does rage within me, but out of concern for the wellbeing of my ship and her crew. They are of a rough-and-tumble disposition, to be sure, and I suspect that more than one has an eye for the pirating life and to use this voyage as a means of achieving that end. But I still carry responsibilities towards them as master of this vessel, and I take such duty most serious. So I would know what you are about, and would be grateful at last for an answer. Why do you take such a lady to this place?'

When Flynt remained unsure of what to say but certain he could not break Woodes Rogers' confidence, it was Cassie who interjected. 'We search for my father, Captain. A man named Gideon Flynt.'

Johnson looked from her to Flynt. 'The boy's grandsire, the seaman? And what business does he have on New Providence?'

'He is in pursuit of my mother, who was abducted from Edinburgh by a man called Toby Hawke. Do you know of him?'

'I have heard the name but never had the pleasure.'

Cassie's eyes narrowed. 'Believe me, meeting him is no pleasure, Captain.'

'And why did this fellow abduct your mother?'

Cassie paused, raised her head as if in defiance. 'He believes her to be his property.'

Johnson was not fazed by the strength of her gaze. 'And is she?'

'No person is another's property.'

'There are those who would give you argument on that score.'

'Are you one such person?'

The captain ruminated on this as he reached for the port again. He refilled his glass, set the bottle back down carefully but did not drink further. He sat back, rubbed his fingers over his scar before clasping his hands over his midriff and taking a deep breath. 'I could lie to you, madam, and tell you that I have never traded those of your race, but I will not. I have transported Africans to the Americas and returned with tobacco and sugar and cotton. And other races, too, to other ports. Not in this ship, but in others, for I have been at sea for a lifetime. I confess that it didn't trouble my conscience at all, for a time.'

By the set of her mouth, Flynt recognised that Cassie had been ready to condemn him, but those last three words made her reconsider. 'For a time?'

Johnson rubbed his head again. 'You have seen my scar, have you not? Of course you have, for it is not something you would easily miss. It was made by a cutlass, wielded by a Madagascar pirate who attacked the ship on which I sailed. We carried men and women, families, from the subcontinent destined for the block in the Dutch colonies on the Cape of Good Hope. The brigands were more interested in the silks and spices we also carried, but they took some of the human cargo to sell at market. They must have had limited space below decks, for they ordered the remainder of the men be tossed overboard to afford their musket sharpshooters the opportunity of target practice. Others made sport with the females. I tried to save a child, torn from his mother's arms – at least I presumed she was his mother for, to my shame, I knew nothing of these people. The child was perhaps no more than two years of age.'

He paused and swallowed, his nose curling as though he tasted something unsavoury, but Flynt recognised the sign of an unpleasant memory. He had many of his own.

'I fought to free the lad,' Johnson said, 'but the fellow who held him was most proficient with his cutlass and he did his best to cleave my skull. Had I not managed to parry the stroke and jerk away, he might have succeeded. As it was, he opened my scalp to such a degree that I was left insensate. He must have thought me dead for he went about his foul business.'

The captain stopped talking, his eyes fixed on his glass again. He seemed to be lost in the memory.

'When I once again woke, I was aboard a longboat with those shipmates who had survived the onslaught. The ship was afire to our stern and from it I could hear the screams of men and women, for they had not freed all of the human cargo. Of the fate of the child I know not, for none of my shipmates could tell me. But it was in that moment of seeing the boy's face creased with terror, hearing his screams, seeing him reach towards the woman as she herself was dragged away, that they became more than just cargo, they became people. It is my belief that the opening of my scalp let some semblance of humanity into my brain. I have never traded in flesh again, and I never will.'

He looked directly into Cassie's eyes.

'Does that make my position clear to you, madam?'

Cassie studied him for a moment, perhaps searching for sign of a lie, but then nodded. 'It does.'

Flynt knew now why the Admiral had wished him to take the *Sprite* and had urged him to make haste in his preparations.

Without a further word, Johnson rose from the table and vanished into his cabin, just off the small wardroom in which they sat. Flynt glanced at Cain, who had reached out to pat Cassie's hand. Flynt had never discussed the matter of slavery and race with his old friend, but he guessed that like many others Cain had never thought much of the right and wrong of it. It was simply a fact of life to most white people, the way of the world. However, Cain's time with Mercy and Cassie would have altered that perspective, of that he was sure. Cain could kill without compunction, but Flynt knew him to be a loyal friend and that would have extended to his family. On seeing the reassuring touch, he once again wondered whether there had indeed been something more than friendship between them.

When Captain Johnson returned he bore a small calfskin wallet tied with a red ribbon, which he placed on the table before Flynt. 'My employer entrusted these with me before we sailed, with instructions to give them to you at the earliest opportunity. That opportunity did not arise until now.'

Flynt peeled back the soft calfskin and lifted the papers within. There were three documents, two being certificates of emancipation for Mercy and Cassie, drawn up by a lawyer of the Admiral's acquaintance. They were as yet unsigned. As he read them, the words they had exchanged in that room above the tavern in Wapping returned to him.

Ship of Thieves

What do you propose to do with them, Jonas?
I'll convince Hawke to sign them.
And if he will not?
Then he will be dead and it won't matter.

15

Flynt stood at the bow of the *Sprite*, the foc's'le deck, he'd come to learn, thanks to Barbecue, who was a veritable fount of knowledge regarding the sea and the vessels that took to the waves. His father and his grandfather and his father before him had all been men of the salt, he'd said, sailing out of Bristol.

'My grandsire lost him a leg to the Dutch fleet at Texel,' the lad had told him. 'It didn't stop him from hopping around none, though, for he was right proficient with a crutch, but I never liked the look of it. It weren't right for a man to be only half of himself.' He shuddered at the memory. 'But he died in his bed of a ripe old age, lacking a pin, but always a-wishing he were back at sea. That be why I'm here, for I loved him, and my old dad. It's a way of honouring them, you see?'

This night, though, Barbecue was nowhere to be seen, and for that Flynt was sorry, for he found the lad's stories entertaining and his conversation informative. So he stood alone, the sun having been swallowed by the darkening sea, the sky losing its colour and the first stars burning through the blackness. The smoking lamp was lit but none of the hands were nearby, though it was certain that they would not be far from needing a pipe or two. Smoking was banned anywhere else on the ship other than around the oil lamp beside the fo'c'sle, a barrel of sand nearby to douse any mishaps. Flame and tarred wood did not make pleasant company, and fear of fire at sea was a pervasive one. The cook's galley below was the sole open fire allowed, and it was constrained by brick and pipe, and only the Irish cook and his assistant – Barbecue – were allowed to enter the cramped space. It was as well that Barbecue was a slim youth, for the cook's full belly was testament to him perhaps sampling too much of his own food. The cooking fire itself was encased in brick with a funnel through the hull to allow the smoke to escape. Flynt was grateful to observe that immense care was taken to ensure that the flame, forever alive, was kept under control.

The evening was warm and he wondered how life on board had changed in just two weeks. On embarkation the crew had fought off the bite of London's cold by clustering round buckets containing heated cannon balls, their bodies wrapped in thick coats. Now, some weeks into the voyage, the coats had been discarded, the ordnance had been returned to the store, and the men went about their business in loose-fitting linen shirts and baggy trousers that ended just below the knee, their calves in woollen socks, although Flynt suspected that even they would be dispensed with as the temperature rose further. Some wore flat shoes, others were barefoot.

Now that he had acquired his sea legs, he found life on board ship peaceful, the warmer breezes and clean air most comforting. There was something about the roll of the keel as it crested the waves, the creak of the masts and the billowing of the canvas as it caught the winds that he had come to appreciate. By day the decks were a bustle of activity as the hands busied themselves with their tasks, climbing the shrouds with ease, clinging to the spars as they furled and unfurled the sheets as instructed, swabbing the decks, tending to the halyards that hoisted the sails and ensuring the tackle that assisted that activity was in good repair. Flynt had learned a number of the terms of the ship from young Barbecue. It had passed the time, and Flynt was always in search of knowledge, though the intricacies of navigation held little fascination.

After dark, there was much less activity. There were still watches to be observed, each one announced by the tolling of the ship's bell. Most of the crew were below, eating, gambling, mending clothes, sleeping, the air ripe with all manner of sounds only a human can make: snoring, laughing, coughing, spitting, farting. Flynt enjoyed these periods of calm as often as he could. He could be gregarious when he had to be – perhaps not as much as Cain, enough to be social is all – and he had thrown the bones and dealt the flats with the crew and had enjoyed himself. However, there were times when his work and his own nature demanded solitude, and he had found it here, on the *Sprite*.

He breathed in the air, fresh and warm, as the waves lapped against the hull, no longer lunging against it as they had before. The ocean vanished into the night in a seemingly endless series of undulations. It was easy to believe that those on board were the only souls left on the earth, so great was the sense of isolation. There were other vessels plying their trades; Flynt had seen them in the daytime, on the horizon or sailing close to, and at night had spotted the burning of a lamp in the dark, a light that was snuffed out by distance or the curve of the earth. The *Sprite* and the ocean were like lovers, sometimes fiery and at odds, but mostly in accord, each needing the other and coexisting peacefully. Was this how Gideon felt when at sea? Was it this sense of freedom that had drawn him away from Edinburgh time and time again?

Footsteps on the short flight of stairs behind him made him turn to find Cassie approaching. Beyond her, on the quarterdeck, young Jonas was being introduced to the mysteries of the steering wheel. Upon a word from his captain, the helmsman stepped away and Johnson motioned for the lad to take the helm. Even from the length of the ship, Flynt could see his broad smile and something bubbled within him, though he knew not what it was. Pride that he had grown into a tall, strong young man? Or was it fear?

'We shouldn't have allowed him to come,' Cassie said.

'I know,' Flynt said, turning away again to face forward. 'But we had little choice.'

Cassie nodded her agreement and they stood in silence for a moment before she said, 'You haven't asked me if Jonas knows.'

Flynt had no need to inquire as to her meaning. 'Given the antagonism he displays towards me, he either knows or at least suspects.'

'Can you blame him?'

'No.'

It must have been difficult for the boy to learn that the man he'd known as his father was not the man who sired him. Flynt had been much older than young Jonas when he had learned that Gideon was not his father, and the shock had been palpable. Even so, he still considered Gideon to be his father and thought himself more Flynt than Moncrieff, which his half-brother, the current Lord Moncrieff, could not understand.

'So he does know?' Flynt asked.

'Aye. He began to suspect when you met, the day you told us that Rab had been killed. He's a clever lad. He maintained his silence for around a twelvemonth then finally made inquiry. I hadn't the heart to keep it from him. Even so, hearing the truth, that the man whose blood he shared was absent from his life not through necessity but by choice, wasn't easy to take.'

Flynt hid the sting of her words by shooting another glance over his shoulder at Jonas standing behind the wheel, both hands on the spokes, still smiling.

Cassie followed his gaze and when she spoke there was regret present. 'I've done my best to ensure he attends to his studies, and he is most attentive, but there is a part of him that yearns for something more than books and laws.'

Flynt had seen it in the boy's expression during the chase in London, and again when he asked Captain Johnson about pirates. It was a spark of excitement, a flicker of enjoyment that he recognised well, for it lived within him, too. Hearing that Gideon had spoken to the boy of his life at sea came as no surprise, for he had done the same with Flynt whenever he returned home, with talk of pirates, daring feats and even sea monsters. Even in his youth, Flynt was forced to consider whether the man he thought of as his father was something more than an ordinary mariner. Had Gideon passed that lust for excitement on to the boy, just as he had with him?

'I need you to talk to him,' Cassie said.

'I suspect he will not relish that.'

She accepted that with a tilt of her head. 'Maybe so, but I need you to try. He must be diverted from the path you took. Gideon has filled his head with romantic tales of foreign lands and adventure. You must show him the reality.'

The way she looked at him conveyed the impression that she wished him to be the father her son needed. Flynt had no idea how he would broach such a thing with the boy, but saying so would not convince Cassie that he was not the man for the job.

'Where is Cain?' Flynt asked.

'He went below to play dice with the crew.'

'He and Jonas have bonded?'

A wry smile compressed Cassie's lips. 'Yes. Gabriel is very charming, but he has not assisted in this regard. Like Gideon, he can tell a grand tale.'

Jealousy stabbed at his chest, and yet he had even less right to that feeling regarding young Jonas than he had over his old friend growing close to Cassie. Even though Cain had denied there being anything more than friendship, doubt still gnawed.

'And you and he have grown close, too?'

She adjusted her head to stare straight at him. 'Yes, but not in the way you suspect. And even if it were, you would have nothing to say about it.'

It was a reproach he could only accept. 'You allowed him to teach the boy how to shoot?'

'It was done at first without my knowledge.'

'And he has a proficiency?'

'Workmanlike. Perhaps not as skilled as his father, but then,' another barb was incoming, Flynt could feel it, 'I believe Jonas lacks that kind of ruthlessness.'

Annoyance with her constant need to rebuke him, however merited, sharpened his tone. 'Let's hope he doesn't have to put that to the test. On that, is there any way I can convince you to remain in Kingston when we dock? You and Jonas, with Cain to watch over you?'

'I need no one to watch over me.'

He felt his exasperation build. 'Cassie, we are heading into a different world, where people like you are seen as commodities...'

'You think that is something I don't know?'

'I'm aware that you know of it, but I feel I have to say this. Stay in Kingston, let me continue to New Providence.'

'Out of the question, Jonas. They have my mother...'

'Yes, I...'

'...and Gideon is as much my father as he is yours.'

He had known she would refuse to see sense, but he believed he had to at least make an attempt. He now thought of the third document that the Admiral had provided, trying to concoct a delicate way of preparing her for it.

'Then we have to take precautions,' he said, pausing, still unsure how to proceed. She waited for him to speak again. 'You saw the bills of sale I've had prepared, awaiting Hawke's signature.'

'He will not sign.'

Then he will be dead...

'But there was another that you didn't see.' He hesitated again and, again, she waited. He sighed and reached under his shirt to produce the calfskin wallet. He thumbed through the thin sheaf of papers and produced the one of which he spoke. He held it out to her. She took it, peered at the writing, then tutted and moved closer to the smoking lamp to cast further light. The illumination it provided was dim and she had to hold the paper close to decipher the words. Once done, her head jerked up sharply.

'You cannot be serious, Jonas...'

'It means nothing in reality...'

'I won't agree...'

'You must, it's for your own protection.'

She read the paper again. Flynt understood her anger, for it was a document that stated she was his property.

'I am to be slave again?'

'No, it's merely a device to protect you from being taken by slave catchers. Should we be separated then you would show them that document and, if luck is with us, you would be unmolested.' He could tell she remained unconvinced. 'You heard the captain, Cassie. You are nothing more to these people than a possession, to some perhaps even less than that. This at least may afford some protection until we can fulfil our purpose and return home.'

'There has to be another way.'

'I can think of none. It's to be hoped we never need it but we have it, just in case. I didn't come to this solution easily, Cassie. I know what it means to you...'

He immediately regretted saying that when she rounded on him. 'How can you? You've never known how it felt to be the property of another, to not have the freedom to even think for yourself, to be regarded lower than the domestic animals. Even as a child I was aware of it, of this overwhelming sense of injustice that I was not seen as a person, that I was merely a... a thing, to be bartered, to be bred from, to be beaten. And killed if the master had such a mind, and for which there was no recourse in law.' Her anger was diffused and she leaned on the gunwale to stare ahead into the smooth darkness. 'That devil Toby Hawke treated his horses better than he did his slaves. We had a friend, Hannibal was his slave name, he never told us the one with which he was born because to repeat it only brought Hawke's ire down on us. He worked in the stables, tending the horses. Hawke was very proud of those animals, the best on the island, he said. But Hannibal made a mistake one day. Hawke had been out riding and the horse was in a sweat. Hannibal was supposed to wipe him down, but he forgot. Hawke discovered this and flew into a temper. He had this walking stick, a heavy, gnarled stick he'd brought from England, and he beat him to death right there in the stables. Nobody could, or would, lift a finger to help. I was outside and I could hear the blows.' She closed her eyes. 'I can still hear them, and Hannibal's cries turning to a whimper and then nothing at all. But the blows kept falling.' She stopped, her head cocked, as if she could actually hear the echo of the solid wood striking flesh. Her eyes opened again. 'They took Hannibal's body and buried it somewhere, we didn't know where. And there were no questions asked. None at all. It was as if Hannibal never existed. No laws protected our kind.'

Flynt took the document from her hand. 'Cassie, this carries no legal weight. It's a ruse, a camouflage, but one I believe is necessary.'

She looked away and in the glint of the lamp he saw the sheen of tears in her eyes. 'I know, but that's not the point. It's... merely the thought of it. I believed we had escaped all that, thanks to Gideon. Mercy believed it, and for the first time in our lives we were truly happy and at peace, even though there were many in Edinburgh who looked askance at us because of the colour of our skin. But we were free. And now, even as part of a ruse, I must take on the mantle of slave once more.'

He folded the certificate of ownership back into the wallet, which he then slid under his shirt once more. 'This will be the end of it, Cassie, I promise you. By the time our work is done you and Mercy will have no further fears.'

When she looked back, for the first time he saw fear in her eyes. Cassie had always been so assured, so certain of herself, so this look took him aback. 'How can you be certain, Jonas?'

He took a moment before he rested both hands on her shoulders. She didn't shrug him off, as he had expected. As he had feared. 'Because I won't allow it. I owe you a great deal, Cassie. You, Mother Mercy, young Jonas. I make you a solemn promise that I will bring this to a satisfactory conclusion for us all.'

He hoped he sounded more confident than he felt.

16

When Jonas came to him two nights later, Flynt suspected that Cassie had encouraged him when Flynt had made no move to do so. His suspicion was proved by the boy's first words.

'Mother says you wish to speak with me.'

He held himself stiffly, his head high, his jaw tense, defiance seeping from his gaze. He seemed ready to lash out, to argue, to blame.

Flynt had no idea what he was going to say, so he continued staring out at the sun as it sank into the waves, making fire of the water. They stood quietly for a few moments before he decided to press right to the heart of the matter. 'You know Rab was not your natural father, correct?'

'Aye.' A catch in Jonas's throat caused a hitch in the single word. 'I know you deserted my mother before I was born and he had a decency you did not.'

Yes, there was the aggression.

Flynt could have argued that when he left he hadn't known Cassie was with child, but he didn't. There was little point in putting up a defence. 'Rab was a good man. He had always admired Cassie... your mother.'

'He was a good father.'

'And a good friend.'

'And yet you let him die.'

Flynt saw sleet darting across a windswept hill. He heard the pistol shot. He saw Rab fall again. He hadn't seen it coming, but he should have. He couldn't defend himself against that charge either.

'I'm sorry, Jonas,' he said, still not looking at the boy.

'Aye, perhaps you are, but if you think that now you will be father to me, then I will save you the trouble. I don't need you. I've never needed you. I can look after my mother without your help.'

Cassie was well able to take care of herself, but that was not the issue at that moment. Flynt now faced him. 'You're a man now, eh?'

'Yes.'

'You think being able to load and point a pistol makes you a man?'

'Do you?'

That caught Flynt unawares and he floundered briefly. 'No, it doesn't. But your mother fears you may believe so. She worries that you see all this as a great adventure.'

'She needn't worry. I know this is a grave business.'

A light that kindled in his eyes proved the reverse. 'I need a favour from you,' Flynt said.

'You need a favour from me?'

'Aye.'

Jonas laughed. 'Ask it then.'

'I need you to convince your mother to remain in Kingston.' This was not what Cassie intended him to do, but Flynt was certain it was the right course of action. 'Gabriel will also remain to protect you both.' He saw an objection forming, so he raised his hand to prevent it from taking shape. 'I know, you need no one to protect you, but he will stay at your side all the same.'

'I wish to help find Grandfather.'

'And you will. There is a chance that he is still on Jamaica or one of the other islands. I would have you make inquiry there, while I move on to New Providence.'

'We know that is where he heads...'

'He may not have reached there, and if we separate then we can cover more ground.'

Flynt was determined to do what he could to keep them safe and, even though he would have preferred to have Cain at his side, he was used to working alone.

Jonas examined this thought and found it made sense. 'My mother will be difficult to convince.'

'I know. She won't listen to me, so it falls to you. She will pay heed to you, I believe.'

'And if she doesn't?'

'Then I will rely on you to protect her, wherever we go. I saw you load the pistol. You must practise more, but you were accomplished. I saw you ready to take action in that lane in London and I understand you were most energetic in Edinburgh. You must never leave her side. I have faith in you, Jonas.'

The boy attempted to maintain his stiff demeanour but Flynt's words, softly but sincerely spoken, weakened his resolve. He couldn't stare Flynt down any longer and his gaze dropped to the deck as if he was searching for something. 'Did you love her?'

'Yes.'

'Do you love her still?'

He thought of Belle. He thought of the words he was ready to say as they danced in London. He thought of how he felt when he saw Cassie standing in that little room in Mother Grady's. 'Yes.'

'When we return to Edinburgh, will you come with us?'

Flynt blinked, hearing in Jonas's voice – convincing himself he heard – a faint plea. 'I don't know.'

Jonas nodded, expecting such a response. 'It's that woman, isn't it? The one in London. The whore.'

Flynt wanted to chastise him for being so blunt but this was not the time. 'No. It's my life. It's not conducive to settling in Edinburgh. But also Cassie, your mother, doesn't want me. Whatever there was before, it's gone now, and that's my fault. She loved Rab, who was the best father you could ever have. She loves you. Neither of you need the baggage that I would bring. I'm sorry for my failings in the past, I truly am, but I cannot make up for them. All I can do is assure you that, should you ever need me, either of you, then I will be there.'

Tears brimmed in Jonas's eyes. He was struggling to maintain his distance but Flynt could see that somehow his words had reached him. He swallowed hard as he held out a hand. Jonas stared at it for what seemed like an eternity, leading him to believe that he was going to reject it, but then he reached out and grasped it firmly. Flynt had the urge to pull him in for an embrace but he didn't. He was not by nature one for such displays of emotion.

When their handshake broke and Jonas left him, even giving him a brief smile, Flynt regretted not holding the boy close, even for a second.

—

The first he knew of the ship approaching three weeks later was a sudden increase in the hustle of the men around him. He was resting in his hammock, reading a book borrowed from the captain's small library, and he swung himself free as Barbecue dashed from the galley towards the ladder leading to the upper deck.

'Sail sighted,' the boy explained as he rushed past. 'Cap'n wants us at the ready.'

'Why?'

'Might be nothin',' Barbecue said over his shoulder, 'might be somethin'. But in these here waters it's best to be prepared.'

Around him, the men unhooked their hammocks and rolled them tightly before darting up the ladder and through the hatch. Flynt followed them into the open, and joined Captain Johnson on the quarterdeck as he peered through his spyglass to the stern. The sun was in the east so Flynt had to shade his eyes against it but he could just make out the shimmering silhouette of the ship. Around him, the hands strapped their rolled-up hammocks against the ship's rails as means of blocking musket fire. The gunports had been opened and the gun crews prepared the cannon for use.

'Do you fear an attack, Captain Johnson?'

The captain lowered his spyglass and looked from Flynt to Cain, who was climbing the ladder to the quarterdeck. 'We're too near the Caribbean to be lax in our security, Mr Flynt. I cannot yet make out that vessel's colours clearly but I will not allow us to be caught with our breeches down.' He stepped around them to call to his first mate, who directed operations on the deck. 'Mr Lamb, raise our colours, if you please.'

The mate nodded once and headed below as the captain resumed his study of the ship. Flynt squinted against the bright sunlight. 'What makes you fear this may be different from all the other vessels that have crossed our path?'

'My gut, sir, and it has served me well these fifty years. I see no colours flying and I sense that is no peaceful merchantman, but a raider. And, damn me, she is closing fast.' He glanced up at the sails, then back aft to the pursuer. 'We have full sail but we won't outrun her.' He regarded Flynt and Cain. 'I would arm yourselves, gentlemen. If I am correct in my assessment then we may need your expertise.'

No further instruction was necessary. They rushed below to where their belongings were stored and fetched their weapons, saying nothing to one another as they returned topsides, where Cassie and young Jonas now waited, observing the flurry of movement on deck with interest. Cassie gave Flynt a querying look but before he could reply Captain Johnson yelled down from the quarterdeck.

'Mistress Gow, it might be advisable that you remain below. We know not what the men aboard that ship are about, but it's best not to let them see you. Our colours will afford us some protection but the sight of you may prove to be too much of a temptation, begging your pardon.'

Cassie was minded to defy him but Flynt touched her elbow and shook his head. 'Cassie, please, remain in your cabin until we understand the situation more fully. The captain is not one to make such a suggestion without foundation.'

Jonas had craned over the gunwale to scrutinise the ship closing in fast, excitement in his voice. 'It's a pirate vessel?'

Captain Johnson, who had resumed his study of the ship, overheard the boy as he lowered his spyglass. 'Aye, lad, she is pirate right enough, I see clear now that she flies the black, plain as day, though I still fail to make out the markings.'

Flynt handed Tact to Jonas. 'Take this and remember what I said before. Remain at your mother's side. Bolt the cabin door, and if someone attempts entrance without knocking, then use it.'

'I would remain here,' the boy protested.

Flynt's exasperation rose. Had he not listened at all? However, he forced it down, knowing that he may well have been of a similar argumentative nature at that age. Indeed, he probably was. Cain must have recognised the flash of anger in Flynt's eye, for he stepped between them.

'Do it, Jonas,' he said, his voice low as he turned the boy slightly away from Cassie. 'Your mother will need your protection.'

He had not spoken softly enough, for Cassie heard, though she didn't react, knowing that the words were spoken only as a means to convince her son. Flynt led her away a few paces. 'Do you have the pistol charged?'

She nodded.

He knew she would. 'In the cabin?'

Another nod. 'You know I am no gentlelady,' she said. 'I can fetch it and be an asset if there is a fight.'

'I need the boy out of harm's way, Cassie. He will not go without you.'

He was using the love between mother and son to keep them safe. He was manipulating them and he felt no guilt. It had to be done. Thankfully, he needed to say nothing further, for she immediately gestured to her son. 'Jonas, come with me. This is not the place for us.'

'Mother, I would rather remain.'

'And I would rather you attend to me, with no further arguments, if you please.' Her voice was sharp but then she blunted it. 'I need you with me and the captain is correct, my presence might inflame matters. Come, smartly now.' She didn't wait for her demand to be tested again, but turned and headed across the deck to the ladder leading to the cabin forward. Jonas was torn between his desire not to miss real-life adventure and his need to protect his mother – and, given the look he shot at Flynt, perhaps fully aware of the game that was being played – but followed her, his shoulders drooping.

Cain waited until they had gone before saying, 'If this does turn bloody, two pistols won't help them much, you know that.'

'I do know that,' Flynt replied, casting his eye at the black and yellow flag now fluttering from the mainmast of the *Sprite*, 'but it's best they stay below until we know how the dice land.' He returned to the quarterdeck. 'Captain, the new colours you fly, I've never seen them before.'

'They are the colours of my employer.'

'You said they might afford us some protection. In what way?'

The captain gave Flynt a steady look. 'You know what manner of man he is. He has many business connections across the seas, including Nassau. Those colours inform whoever is on yonder ship who owns this vessel.'

The Admiral had an arrangement with the pirates of New Providence. That was interesting. It was to be hoped that his reputation was as potent in these blue waters as it was in the murky anchorages on the Thames.

Mr Lamb had maintained a close eye on the ship through his own spyglass and leaned in closer to murmur something. Captain Johnson swore softly and jerked his glass to his eye to study the ship anew.

'You're right, by God, those are his colours,' he said, sighing heavily, then turning to Flynt, his expression grim.

'I take it this is not good news,' Cain said.

'Had it been any one of a number of these rogues then I might have taken my ease over this encounter. Scoundrels and wastrels they be, but many are not unreasonable men and they would have honoured the Admiral's colours. But perhaps not the one who commands this particular vessel.'

'Who is he?'

The captain handed him the telescope. 'Focus on the mainmast, on the flag.'

Flynt did as he was told. The fluttering black cloth bore the image of what appeared to be some form of demon, or at least the skeleton of a horned man, with an hourglass in its right hand and a spear in its left about to pierce a blood-red heart.

'Each pirate his own colours, some black, some red, most bearing an image, such as that one.'

Flynt handed the glass to Gabriel. 'So whose ship is that?'

The captain's face was grim. 'The captain of that ship is the one they call Blackbeard…'

17

Sight of the flag caused deeper unease to flow around the crew. Flynt could feel its current surging around him, and saw it in the way they cast anxious glances at the ship as it drew closer. They continued to make their preparations but with considerably more expedition. Previously the threat had been something vague. There was a ship, to be sure, and it may have been pirate, but now that ship had a captain and a reputation that preceded him.

Blackbeard. Flynt had heard the name before Charters had made mention, but as he said, only in passing. A mention in a dockside tavern, a suggestion that he was without compassion, perhaps more demon than man. Stories, whispered over ale and brandy and fuelled by both. Of how he showed no quarter, of how he lit tapers in the pleats of his long facial hair and set them alight, of how syphilis had made him mad. Of how he killed without compunction.

Flynt and Cain double-checked their pistols, Flynt also twisting the handle of his silver cane to unlock the sword blade hidden within, then thrusting it into his belt like a scabbard. A wry smile played on Cain's lips.

'I'll wager you wish you had retained your pistol, eh, Jonas? If this comes to a fight then you might need that second ball.'

'If this comes to a fight, then a second ball won't make much difference.'

Once again, Captain Johnson overheard them. 'Make no move towards aggression unless I order it, gentlemen. These are not the gutters nor heaths of London, and you are not the masters here. You're under my command and you will obey.'

'Aye, aye, Captain,' Cain said, with a quick salute, then dropped his voice. 'I swear that man has a gift of hearing that borders on the supernatural.'

Flynt made no reply, for he saw the captain give Cain a sharp look, then a brisk shake of his head. He had spoken barely above a whisper and yet Johnson had heard him. Perhaps there was indeed some skill that was not of this world in play.

Nothing further was said as they waited for the ship to gain on them. The gun crews stood nervously at their posts; other men hid below the level of the gunwales armed with muskets, pistols and cutlasses. A few had climbed the standing rigging, weapons slung over their backs, to find a vantage point from which to fire. This was a merchant ship but one that was used to the fight. Nervous they may have been, but the men aboard were ready to defend themselves.

The approaching ship was larger than the *Sprite*. Its masts and full sails towered over them, its decks riding higher in the water, casting full shadow over the smaller ship. As it neared, the gunport on its starboard side opened and the figures of men stood ready on deck or swarmed up the shrouds. There was silence for a few moments, broken only by the creaking of the timbers and the flap of sails in the wind. Then a voice floated across the water between them.

'Why the preparations for battle, lads?' The voice was coarse, the accent west country. 'We mean you no harm.'

Mr Lamb handed the captain a copper tube, wider at one end. He placed the narrow opening to his lips and shouted into it. 'And yet you come upon us with open ports and men at the ready.'

'Because we perceive you lads doing similar. We sees your colours and wish no friction with the Admiral. He be friend to our brethren.'

Flynt tried to make out the owner of the voice, who either used a speaking trumpet similar to the captain's or he had extraordinarily powerful lungs, but he couldn't pick him out among the faces opposite. He presumed him to be on the quarterdeck, though.

'Then what do you seek?' Johnson demanded. 'You come upon us in full sail flying the black, so what do you expect any law-abiding vessel to think?'

There was a pause in the exchange, and then a laugh floated towards them. 'Aye, I reckon you would take such precautions, and that be right and proper, for you'll no doubt know who is master of this here vessel. And who might I be having the pleasure of addressing, if it ain't too bold to be asking?'

'Captain Charles Johnson, master of the *Sprite*, the owner of which you are already aware. And do I address Edward Thatch?'

'No, the captain is below. He has entrusted this here transaction to me.'

'Then who are you, sir?'

'I'm Hesikia Hands by name, some might call me Basilica, quartermaster on the *Queen Anne's Revenge*, as she be now, formerly *La Concorde*. But I heard tell of a Charlie Johnson what former sailed with that fool Woodes Rogers.'

'I had the honour of being privateer with Captain Rogers, sailing by lawful letter of marque in service of King and the laws of England and, mark me, I will not be cowed by your show of force. If you attempt to board us, my men and I will fight you to the end, blood and bone.'

Another laugh. 'Well spoke, Cap'n dear, but I say again we mean you and your brave men no harm.'

'Then I say again, what do you seek?'

'Not your ship nor your cargo, on that you has my oath.'

'Such as that oath is,' the captain breathed, then placed the trumpet horn to his lips once more. 'What then? Speak, man, for I am eager to reach Kingston.'

'I'll say it plain, for I sense that you is a man what likes straight speaking. We need men, in short, for our last engagement left us powerful wanting of crew.'

'If that be them short of men, I'd be loath to see them at full complement,' Cain said and Flynt had to agree.

'My men are not desirous of the life of pirate, Mr Hands,' the captain said.

'That be what many a captain says, but oftentimes their men takes themselves a differing view. Why not allow me to put it to them?'

'You will coerce them, Mr Hands?'

Another laugh floated their way. 'I will make a propositionary argument, Cap'n dear.'

The captain frowned and turned away, as if he feared the man on the galleon opposite would divine his words from his lips. 'We are in something of a pickle, gentlemen.'

'How many men do you think will desert the ship to join them?'

'What say you, Mr Lamb?'

The first mate agreed. 'I can think of three most immediate who would elect to turn freebooter.'

The captain nodded. 'Aye. And even three would leave us short-handed for the remainder of the voyage, but if I refuse him access, this man Hands might take it upon himself to unload those guns.'

Cain asked, 'In that case, why not get your retaliation in first?'

'My men are hand-picked and have all seen action of some sort, to be sure. We may have luck and bring down a mast but the fact remains that we are outgunned, sir, and let me assure you the *Sprite* cannot withstand a broadside of such magnitude.'

'But the man Hands assured you that they had no ill intent towards the ship.'

'The man is a pirate, Mr Cain, and the truth is a foreign language to such men.' Captain Johnson faced Flynt. 'But it occurs to me that there might be something providential in this situation. You need to reach Nassau, and I will lay you any odds you care to name that they are returning there with that ship's hold full of stolen goods. They have seen action and lost men, so if you were to offer your services to them, it would be a means of reaching your destination far swifter than you will if you remain with the *Sprite*. From this latitude you would reach the island in less than a week, but if you remain with us it will take considerably longer.'

Flynt saw a means of separating himself from Cassie and Jonas. But then an impediment raised in his mind.

'What can I offer them that would be of use? I have no knowledge of the sea, navigation nor how to handle a ship.'

'It's true,' said Cain, 'he can't even tie a decent knot. He believes a sheep shank to be a cut of meat.'

'There is more to manning such a vessel than seamanship,' said Captain Johnson. 'Pray leave it to me...'

—

Hesikia Hands arrived on board with two men only, who had rowed across the divide in a jolly boat, hauling at the oars with strong, confident strokes, while he sat aft, manning the tiller. They were all well-armed, each with a brace of pistols tucked into a wide belt from which leather scabbards hung carrying their cutlasses, though one favoured a longer sword that was even more curved, and they had muskets slung across their backs. Hands was a powerfully built man who towered over his more diminutive companions, who though lean were firmly muscled. Their skins were sun-darkened and their hair tied back by brightly coloured ribbons. Their silk shirts, one red, the other yellow, were of fine quality – Flynt suspected the garments had been liberated from a plundered ship – and their taffeta breeches were black. Their calves were bare of hose and they wore black buckled shoes showing the worse for wear. Salt spray was perhaps not good for the leather. Hands himself wore no shirt but did have a waistcoat over his muscled torso and a leather pouch draped across his chest. Flynt would lay odds that it carried cast-iron grenades. Hands had claimed to come in peace, but he was prepared for action.

After they had hauled themselves up the side ladder and onto the deck, Hands ascended immediately to the quarterdeck and graciously thanked Captain Johnson for the opportunity to speak to his men.

'No gratitude is necessary, Mr Hands,' the captain replied, somewhat testily, 'for I suspect I had little choice in the matter.'

Hands grinned, showing an array of discoloured teeth, some blackening from the root. 'We all has us our choices, Cap'n dear, and it speaks well of you that you be giving your men their opportunity to make their'n.'

The captain waved his hand to the assembled men on the deck below. 'Then make your case, Mr Hands, for I am eager to be underway once more.'

'I won't leave you so lacking in manpower that you will be unable to sail this here vessel, on that you has my solemn oath.'

'Then you must proceed with all haste.'

'Thank you kindly, Cap'n dear.' Hands stepped to the rail to gaze down on the faces of the men below. 'Now, lads, there ain't no point in my dragging anchor over this, for you will all have heard the exchange between me and your fine cap'n here. You be knowledgeable who I be and who my cap'n be. You will knows what I wants: good men, solid men, men who knows their way around powder and shot, but who can stand firm in the teeth of a nor'wester and who ain't afeared to do so.' He made another survey of the upturned faces. 'Standing here I sees some right sturdy lads, and would be proper proud to have you as shipmates, so anyone what wants to live a free life should step forward now and we'll take you back to the *Revenge* smartish.'

Nobody moved, but one voice carried from the rear, a Welsh accent. 'And what if we don't choose? What happens to the *Sprite*?'

'I gives my word that this ship will proceed unmolested and I stand by that. We ain't here for no blood, just brawn and guts, if any will give it, free and clear and without impediment. Now, I knows that some of you men will hanker for the kind of life we offer, but I will say up front that it's a hard life, a dangerous one. But then, that is the life of a seaman, am I right?'

There was a murmur of assent and a few heads nodded.

'But I can give you my word that with us the rewards is far greater.'

Captain Johnson cleared his throat. 'Aye, but the only way a pirate ends his days is at the end of a rope, and that time is coming for the men of Nassau, mark me, Mr Hands.'

Hesikia Hands grinned again. 'Aye, p'raps that be the right of it, but similar can occur if you be an honest seaman, ain't I right, Cap'n? Step out of line and there's a flogging, get into a fight with a fellow seaman and kill him inadvertent and it's a short drop off the yardarm for you, similar if you makes a simple question of a course or an order and it be deemed mutiny. Life is hard no matter which ways it is scrutineered, but at least with us a man ends a voyage with a heavier purse to set against his future years, if you is canny and doesn't drink it or wench it all away in the taverns. So what says you, lads? Who among you is ready for a life of adventure and the opportunity to no more bow your head to them what believes them to be our betters?'

Feet shuffled and glances were exchanged for a few moments before first one man shouldered his way to the front, then another and another. Finally, Barbecue raised his hand and eased to the front, avoiding the reproving look from Captain Johnson.

'You're making an error, boy,' he said.

Barbecue still couldn't meet his former captain's eye. 'I've got to be my own man, Cap'n sir.'

Hands laughed. 'That's the spirit, my lad, this will be the making of you, on that you can be sure and certain.'

The pirate quartermaster waited to see if any further men decided they wanted to jump ship, but when none came, he seemed disappointed. 'I admit, free and easy, that I thought there would be more of you, but you is all grown men and you make your own destiny.' He turned to Captain Johnson. 'I'll take these here four now, Cap'n dear, and you can be on your way, as I promised.'

'I implore you, Mr Hands, leave the boy. He knows not his own mind.'

'He looks well able to have the knowledge of it to me. How old are you, lad?'

'Not rightly certain, Mr Hands, but I'm old enough sufficient for the work ahead.'

Hands turned back to the captain. 'He seems sturdy enough for me, Cap'n dear. Ages with my own lad, Israel.'

A surreptitious glance from the captain prompted Flynt to speak. 'Leave the boy and take me.'

Hands stepped nearer to make a close study of him, his gaze moving from head to toe, taking in the pistol in his belt, his eyes narrowing when they fell upon the silver cane. 'You don't look like no salt to me. What use would you be to us, I asks?'

It was Captain Johnson who provided the answer, injecting a grudging tone into his voice. 'He is a fighting man, Mr Hands, on his way to the colonies to make his fortune.'

Hands raised an eyebrow. 'That be so?'

'Something like that,' Flynt said.

Hands detected the tone that Flynt intended. 'Something like that?' He sidled closer. 'You fleeing the law in England, or maybe Scotland, for I hears the heather in your voice.'

Flynt shrugged and Hands' attention switched to Cain, giving him as close a scrutiny. 'And you? Be you a fighting man, too?'

'I've been known to have the occasional disagreement.'

'*The occasional disagreement*,' Hands repeated, walking round them both, coming to a halt at Flynt's back. Flynt could feel his eyes taking in every inch. 'Maybe you is and maybe you isn't fighting men. There be only one way to find out for sure.'

The sound of steel sliding on leather made Flynt move instinctively. He jerked his elbow back and rammed it at a level he hoped Hands' eye rested and was gratified to feel it land. He leaped forward, twisting round as he drew his own blade free. Hands stood with his cutlass in his hand, his other holding his left eye. 'Kind of twitchy, ain't you, m'dear?'

'I become so when I hear a sword being drawn.'

Hands stopped rubbing his eye and waved towards the men on the deck.

'You is fast, I does readily admit, but these here will be true men of the salt, elsewise they wouldn't be aboard this vessel. I can smell the sea off them, even the lad. But you? I doesn't smell it from you. You're a landsman to your very vitals.'

'As Captain Johnson said, I'm no sailor. All I offer is my aim and my arm.'

Hands gestured towards his eye and then jutted his chin towards the sword in Flynt's hand. 'You certainly have the instinct of a fighter, landsman, I'll grant you that. But does you have the stomach for it? That's what my cap'n will wish to know.'

Flynt said nothing, but steeled himself for what was about to happen.

Hands' smile slid away. 'How proficient be you with that there toothpick?'

'I remain standing while others do not.'

'Maybe so, maybe so.' He flicked a finger to one of his companions, the other nowhere to be seen. 'Give him a proper blade. Let's see what manner of fighting man he be...'

18

Used to the delicate balance of his own sword, Flynt found the cutlass somewhat alien. He recognised the need of it, for strong though the German steel of his weapon was, it might not be sufficient to withstand the heavy blows of Hands' weapon. Nevertheless, he toyed with the notion of refusing it and retaining his own, for its lightness might afford him an advantage, but in the end, he decided to accept the weightier blade. He suspected that the offering of it was part of the test and he didn't wish to fail before a blow was struck. He stepped away, flexing his shoulder by swinging the cutlass back and forward, up and down.

Cain followed. 'This fellow Hands looks somewhat formidable,' he said.

'I noticed.'

'Are you sure you're up to this? He seems most at ease with that weapon.'

Flynt glanced back to see Hands handling his own cutlass with considerable dexterity, twirling it round in his hand and slicing the air ahead of him. Flynt recognised it as more a show of expertise than a means of limbering up.

'I don't believe he means to kill me, just to test me.'

Cain watched the display for a second or two. 'I wish I had your faith. Still, look on the bright side, it could be worse.' He smiled. 'It could have been me he challenged.'

Flynt gave him a sour look. 'Thank you for sustaining me in my hour of need.'

'Think nothing of it – that's what friends are for. Now, go and show him what you're made of. Not literally, of course, for that would be most inconvenient.'

'If I emerge victorious, then I entrust Cassie and Jonas to you. Go to Kingston, but delay as long as you can their departure. It's to be hoped that I can find Gideon expeditiously and join you there.'

'Cassie won't welcome that.'

'Use that charm I hear so much about.'

Flynt handed him his remaining pistol and his sword cane while Hands pulled the strap of the leather pouch over his head and entrusted it to one of his men. Again, Flynt considered where the third pirate had vanished to and a small measure of alarm crawled inside his head, but before he could make any form of inquiry, Hands lunged at him and thoughts of the man's location were banished. Flynt leaped back, parrying the stroke with ease. The captain, Mr Lamb and Cain, as one man, moved further away to give them room.

Hands grinned. 'You're fast, landsman, I'll swear to that.'

'You're no sluggard yourself, Mr Hands.'

Flynt jumped, the cutlass swirling around his head and then arcing towards Hands, knowing the man would react swiftly, but wishing to show that he really was not one to linger. Hands' weapon clanged against the blade and swept it away, before he danced a few steps out of reach, his sword arm dropping to his side. He strutted around the quarterdeck, his eyes never leaving Flynt.

'You have to do better than that, m'dear,' he said.

'I'm still warming up,' said Flynt.

'I'm always warm,' Hands replied and closed in again, his sword bearing downwards, but Flynt raised his weapon to bring it to a shuddering halt, the power of the stroke vibrating along his arms. Each man grabbed the other's wrist with their free hand and they twisted round in a curious little dance, each testing the other's strength, before Hands jerked his knee upwards, aiming for Flynt's groin. Thankfully, Flynt twisted his body so the blow landed on his upper thigh. It was painful but not as debilitating as the intended strike would have been. He jutted his head forward, the hard part of his forehead connecting with Hands' nose, sending him staggering, his fingers touching his nostrils to see if there was blood. Seeing none, he surged back again, hacking and slashing with increased vigour. Flynt, his head stinging from the blow he himself had delivered, blocked the attacks, but was forced by their ferocity to retreat. Hands persisted, the sword arcing and swinging and thrusting, the strength of his onslaught considerable, the impact of each blow passing again from the blade to Flynt's arms. The man was strong and he was more than capable with the cutlass, while Flynt was still growing used to it.

As Hands made another onslaught, Flynt neatly stepped to one side and slammed the hilt of his sword into his temple. The quartermaster reeled and Flynt pursued, arcing the flat of his cutlass towards his face, having decided that killing or seriously wounding the man would not suit his purpose. He had to disable him without permanent injury. The blade slapped against Hands' cheek, causing him to lurch away again, but he remained upright. He was made of stern stuff, for he recovered immediately and with a snarl surged back towards Flynt, his thrusts and attacks perhaps not as powerful as before, but still potent. Flynt parried each one, ever vigilant for another opening that he could turn to his advantage, but Hands was doing the same.

Flynt swung, knowing immediately that he had misjudged the timing and speed, allowing Hands to duck under his blade and slice at him, ripping a small gash in his belly. He dabbed at the wound quickly, judged it to be not very deep, even though it bled profusely. It was far from mortal, little more than a nick, but the pain was intense and immediate. Hands struck again, Flynt falling back, unaware that he was perched at the top of the quarterdeck steps. He tumbled, landing heavily on the upper deck, where the onlookers pulled away. The air forcibly expelled from his lungs, an ache from his back joining the bite of the blade on his belly, Flynt just in time saw Hands leap after him, sword raised to strike. He summoned the strength to roll to the side just as the steel slammed into the boards where he'd only moments before lain. Flynt had no intention of harming Hands, but Hands had other ideas.

Lashing out with his right leg to catch the pirate on his left knee, causing his leg to buckle, was a sufficient impediment to allow Flynt to pull himself to his feet and follow up with another kick, taking a leaf out of Hands' own strategy by landing between his legs. The pirate's knees drew together and a high-pitched squeal escaped his lips. Flynt ignored his own pain to press the advantage but Hands managed to deflect his next sword strike, slamming his shoulder into Flynt's chest to push him away. Flynt swung the hilt of his cutlass backwards and rammed it into the man's mouth with as much force as he could muster. A curse escaped the pirate's lips, along with a small spray of blood. They each took the opportunity to put some space between them, but circled one another like wary animals, their breath ragged, sweat seeping from their pores. Flynt adjusted his hold on the cutlass, the moisture on his palm making it slippery.

Hands' mouth worked and he spat out a blob of blood, along with a tooth, which clattered on the deck. He stared at it and sighed at the loss. 'You fight well, for a landsman,' he said, his voice lacerated by his heavy breathing.

Flynt swallowed, took a breath, before replying, 'You acquit yourself not too badly also, for a pirate.'

Hands smiled and Flynt saw the gap where that tooth had once nestled. 'What say we put up our blades, then? I am satisfied that you are indeed a fighting man, and my head is exceeding foggy.'

Flynt was grateful for the suggestion but remained cautious. 'You first.'

Hands grinned once more. 'You can trust me, landsman.'

'Never trust a man who says you can trust him,' said Flynt. 'My father taught me that, and I've learned that it holds true.'

Hands laughed and dropped his cutlass on the deck between them. 'Your father spoke wisely, to be sure.'

Flynt threw his weapon down. 'That he did.'

'You'll do, landsman, you'll do us right nicely.' Hands held out his hand. 'Come, let us shake upon it.'

Flynt detected no subterfuge in Hands' expansive grin, gap-toothed though it now was, and he grasped the man's hand, though he remained tensed for another attack. Hands twisted round to face Captain Johnson, watching from the quarterdeck.

'We'll take this here one, Cap'n, and them others. It won't be enough to make up our full complement, but I'll lay true to my word and that of my cap'n. There won't be no impressment of them that is unwilling to join.' His eye fell on Barbecue. 'And as a further show of good faith, and as a sign of respect to your owner, we'll leave the boy here behind.'

Barbecue protested but Hands raised a hand. 'If you has wishes to join us, lad, then once you make landfall you find your way to Nassau. If you truly be a man of deliberation, then you will show us such proofs. This here landsman took up sword for you and a bargain is a bargain.'

Barbecue seemed about to argue the point but then thought better of it. Flynt was relieved but suspected it was only a temporary respite. The lad sought adventure, and he would have it. In many ways, he reminded him of Jack back in London. Jonas had similar leanings, but he lacked the devil-may-care attitude of those boys.

Around him, Flynt sensed the tension among the *Sprite* crew ease. Had Hands not been a man of his word, there was no doubt that they could have overwhelmed him and his two men with ease, but there remained the threat of the massed guns on the ship alongside. Captain Johnson nodded his satisfaction at the outcome, while Flynt again mused on where Hands' second companion had gone, searching the faces of the men clustered around them for sight of him. Where the hell could the man be?

The solution came with the sound of a pistol shot, prompting Hands to snatch up his cutlass once more, while his remaining man threw him a pistol. The sound had come from the forward compartment and Flynt's heart sank when he realised that Cassie and Jonas had been discovered.

Sure enough, the missing pirate pushed the boy ahead of him, dragging a struggling Cassie behind. She swung a clenched fist in his direction but he ducked and, laughing, pitched her forward onto the deck. Jonas cursed him and launched himself but the pirate was prepared, for he dodged out of his path and toppled him with a blow to the side of the head.

Hands pushed his way through the crew members. 'Who does you have here, Rufus?'

The man called Rufus hauled Cassie back to her feet. 'Found this one in a cabin for'ard, with this here lad. The woman took a shot at me with this pistol.' His accent was curious, nasal and a mix of west country and Scottish. He brandished Cassie's pistol.

'Did she, now?' Hands said, good-humouredly. 'I takes it she missed?'

'By a whisker, Hesikia, by a whisker. Had I not ducked back I wouldn't be standing here now.' He produced Tact from where it was tucked in the back of his belt. 'The lad had this here pistol but he couldn't pull the trigger.'

Jonas had frozen. Flynt had warned him that shooting at a man was different from practice. He wasn't surprised and he wasn't disappointed, though they were lucky that these men were not intent on bloodshed this day.

Hands took the pistol, examined it closely, then glanced in Flynt's direction. 'This here be identical to the barker you have, landsman.'

'That's because it's mine,' Flynt admitted. 'I gave it to him before you boarded, for we had no idea if you were raiders.'

Rufus ran a slow, lingering hand over Cassie's back. 'She is a fine, sturdy piece, Hesikia.'

His fingers moved down to her buttocks, but she snatched herself free of his grasp and swung another blow, which this time connected. Rufus's expression was one of surprise and pain and he stepped back, rubbing his jaw. He seemed in good temper, however.

'She's a spirited one, and I like that in a wench,' he said. 'Means they move most lithe under you in the bunk.'

'Give me a weapon and I'll show you how lithe I can be,' Cassie said.

Hands laughed, then said to Flynt, 'She is slave to you?'

'I am no slave,' Cassie said.

Flynt's mind jumped to the forged papers, and that momentary falter allowed Cain to jump in. 'She is my wife.'

Hands looked up at him, then leaned over to examine Jonas's features. 'The boy don't favour you none.'

Cain blinked, aware that there was no resemblance. 'His father, my wife's first husband, died,' he said, carefully.

Hands gave Jonas another close study, then addressed Cassie. 'You prefer the white meat, girl?'

'I prefer men who are decent and honourable,' Cassie flared at Hands. 'Such descriptions do not apply to you or your crewmates.'

Hands laughed again. 'And what happened to your first husband, I wonder?'

'He died.'

'I trust not at your hand.' He addressed Cain again. 'And she is a free woman?'

'I can speak for myself,' Cassie said. 'I am free, as all God's creatures are meant to be.'

Hands moved around her, his eyes roaming up and down her body. Cassie's tight expression revealed the effort she was making not to turn with him. She would not show fear.

'But you was not always free, I'll wager,' said Hands. 'Your voice is Scotch but I still hear the islands there.'

'I left when I was young.'

Hands stood before her again. 'And now you return. Why would that be?'

'I told you,' Cain said, moving down the quarterdeck steps, one hand on his pistol. 'She is my woman and she goes where I go. We have papers of manumission.'

Hands turned to face him. 'And why do you go to the islands? You're no seaman, that is plain and clear, nor less a planter, and given you has married a Black woman, I reckon you won't be liking what you sees on no plantation.' His eyes flicked to Flynt and then back to Cain. 'You two seem to be like peas in a pod, though one be fair and t'other be dark, but not brothers. I'd wager all my prize shares that you both are fugitives from English law?'

'Let's say that we go to the colonies to make our fortune,' Cain said. 'There are more ways to achieve that than planting tobacco or cotton.'

Hands was thoughtful. 'Aye.' He scrutinised both men, squinted towards Cassie and young Jonas, then swung his sword lazily towards Flynt. 'We'll take you, but you,' he swung the tip towards Cain, 'you stay here and keep your wife with you.'

Rufus did not appreciate that decision and pulled Cassie closer to him. She struggled but he held her firm. Flynt resisted the urge to cross the deck and teach him some manners.

'Now, hold there,' Rufus protested, 'let's talk about this...'

'I won't be taking no female on board the *Revenge*. You knows the regulations. No women, no matter how comely they be.' Rufus's eyes blazed with defiance. 'Do you want to explain to the cap'n why you've breached the articles?'

Rufus sneered. 'The cap'n ain't one to speak. How many wives does he have now? I heard sixteen at last count.'

Hesikia dismissed that with a wave of a hand. 'Any woman is a disruption aboard ship, but a comely one can lead to high passions, and that way lies bloodshed. I won't have no such confusion aboard my ship.'

Cassie was prepared to protest at being so described but Flynt gave her a surreptitious shake of the head and she thankfully remained silent. Rufus was also of a mind to continue his own dissent, but Hands headed it off by saying sharply, 'The order has been given. Don't make me repeat myself.'

Rufus's mouth clamped shut but his anger was evident. Hands gave Jonas another look then returned the pistol to Flynt with a curious gleam in his eye. 'You can fight, landsman, on that I be certain, but there's something about you that don't smell right.'

Flynt accepted the weapon. 'That's been said before.'

'The cap'n will be wishing to have words with you.' Hands raised his voice so the other recruits could hear. 'With you all, I'll be bound, but with you in particular, landsman. What do they call you, anyhow?'

Flynt thought it unwise to reveal his true name, so he gave the first one that he could think of. 'Joseph Blake.'

'Scotchman, am I right?'

'Aye.'

'We have some Scotchmen aboard, so you'll feel right to home. Fetch your traps, we set off immediate for the *Revenge*.'

Cain followed Flynt to the ladder. 'I don't like this.'

'It gets me to Nassau all the quicker.'

Cassie and her son had also joined them by then. 'Without us,' she said. 'This was what you wished, is it not? You'll leave us behind, for that is what you do well, is it not? Leaving people behind.'

It was said that truth hurts and once again her words stabbed at him, and a glance towards Jonas revealed a sliver of his old pugnacity, but this time he hit back with his own sharp tone. 'Yes, it is, but only because time is of the essence, Cassie. Gideon has a start on us and we have no idea how quickly he would reach New Providence. We have been given the means by which at least I can get there expeditiously and we must take it.' It was on the tip of his tongue to say that he was better alone but he knew it would only cause her to lash back at him. He forced a more conciliatory inflection. 'I've already explained to Jonas how useful you and he will be in Kingston.'

Cassie shot a questioning glance at her son, who remained silent. Flynt guessed that something troubled him, and it wasn't him leaving. 'Don't let what happened in the cabin trouble you.'

The boy's eyes revealed shame. 'You were right. When it came to it, I couldn't fire, not at a man, not even *that* man.'

Cain clapped him on the shoulder. 'It happens, boy.'

'Not to you, nor to you, Mother.' Jonas looked again to Flynt as he thrust a clean shirt and breeches into a canvas bag, along with the forged bills of sale. 'And it wouldn't happen to you.'

Flynt tied the drawstrings on the bag and threw it over his shoulder, then faced the boy. 'We have all led different lives.'

'You knew I wouldn't be able to do it.'

Flynt risked laying his free hand on the boy's shoulder, expecting him to pull away, but he didn't. 'There is no shame to it, for it needs those different lives in order to be able to take a life. There's no pride to be found in being able to do so, Jonas.'

Jonas's face stiffened and a hint of that defiance crept back into his eyes. 'Next time I shall do what has to be done.'

They stared at one another for a moment, the man who was no father and the boy who was his son, and Flynt tried to think of something to say that might both comfort him and deflect him away from such a future eventuality. But he found no words.

Instead, he held the fake ownership document out to Cain.

'I take it I am to become Jonas Flynt,' Cain said, with a grin in his eyes.

'The fulfilment of an ambition, I'm sure,' replied Flynt.

Cain grunted. 'You think very highly of yourself.'

Cassie watched the exchange, her disapproval evident as she looked Flynt up and down, but remained silent. She reached out towards Flynt's midriff. 'You're hurt.'

He'd momentarily forgotten about the sword wound but now that it was mentioned he again felt its sting. 'Mr Hands was most adept with his cutlass. He cut just enough to bleed but not to kill.'

She began to examine the gash. 'That must be cleaned.'

He gently gripped her fingers. 'I have not the time.'

Her grimace was one of irritation. 'One day someone will cut deep enough to kill.'

He lifted his wide-brimmed hat and placed it on his head, then his heavy coat from where it hung on a nail and draped it over his arm. 'Not today. And I will endeavour to ensure that it is not for many days yet to come.'

With a final nod to Gabriel, and a lingering look towards Cassie, whose anger had not dissipated, he began to climb the ladder to the deck.

Behind him, he heard Cassie say softly, 'Have a care, Jonas. You are not immortal.'

19

Standing on the deck of the *Revenge* and looking down on the *Sprite* gave Flynt a more complete understanding of the sheer power held by the pirate ship. Captain Johnson had been correct, for if they had engaged, the little merchantman would have come off considerably the worst. An impressive array of cannon lined both sides of the main deck while a lower deck also boasted a line of gunports. The pirate crew milled around him, a far greater complement than the smaller ship, many of the men shooting curious glances in his direction as they passed. He heard one whispered comment between two seamen that he was the man who had very nearly bested their quartermaster. Flynt was aware that such was not quite the case, but in this world, as much as in London, reputation could often be what stood between life and death.

The other men who had elected to join the pirates were taken one by one to a cabin in the stern, presumably to meet Captain Thatch. Hands had explained that he liked to speak to new recruits personally, to get their measure. Flynt was the last to be summoned and so he was able to watch the *Sprite* depart to the south-west on its course to Jamaica, while the *Revenge* continued in a westerly direction for the Bahamian archipelago. The sight of the brigantine's sails slowly receding brought with it an unexpected sense of melancholy. It was true that he worked more efficiently without having to concern himself with the safety of others, but he found himself mourning the opportunity to spend more time not just with Cassie, but also Jonas. And that not only saddened him but surprised him. In his work emotions could lead to lethal errors. He had learned to suppress them, to fasten them down tight, but he carried with him a great deal of guilt. Bad decisions, wrongdoing, loss of friends, betraying Cassie in his youth. Now to that tally he must add Belle. He had convinced himself that he truly loved her – even though he did, in point of fact, love her but only after a fashion. She had spotted in an instant that he was still besotted with Cassie, and his realisation of that only increased his emotional pain, which bit sharper even than the flesh wound left by Hands' blade.

He took a deep breath, sucked in the warm air with its salty tang, let the boom of the breeze catching the sails on the three tall masts fill his hearing. He had a job to do and he would do it. Reach Nassau, find Gideon and Mercy, take them home. He had no idea how he would achieve it but he knew he must try. What he had told Jonas and Cassie was true. His father could still be on Jamaica, or another island, trying to find a way to reach Nassau, but he could just as easily have already reached there to face perils he might not be able to handle. It was imperative that Flynt make landfall as swiftly as he could and this man Thatch and his vessel was a means to that end. Flynt always made the most of such eventualities. Flynt always did what had to be done.

He heard Mr Hands bellowing an order to one of the men about massing the crew on deck, and opened his eyes to see him striding towards him.

'Taking a moment, are we?'

Flynt smiled. 'We all need moments of calm reflection.'

'That we does, landsman, that we does.' He jerked his head to the doorway leading under the quarterdeck. 'Your presence is requested by my cap'n.'

Flynt hefted his sack, hat and coat, which he placed against the rail. Hands watched, then said, 'One thing, a word of advice, or caution, call it what you will. Refer to him as Cap'n Thatch, or Cap'n only, unless he gives you leave otherwise. He don't like the name Blackbeard.'

'Why not?'

Hands shrugged. 'It's something what has been attached to him, thanks to his physicals, and as such it has its uses in our profession, but he doesn't much care for him being addressed by it personal. So Cap'n Thatch, or just Cap'n, and that will suffice.'

Reputation could often be what stands between life and death.

As he followed the quartermaster through the doorway and into a dim corridor that led to a further door, Flynt's earlier thought returned to him. His own notoriety in the streets of London had prevented many an ugly scene. He also disliked his own soubriquet – the Paladin – so understood why this pirate leader was unhappy.

Hands motioned for him to wait, then rapped his knuckles on the door before opening it without waiting for permission, but then, the man beyond that portal was expecting them.

The quartermaster ducked his head round the open doorway. 'The landsman, Cap'n.'

No order to allow him entrance was given, but Hands nodded once then crooked his finger in his direction. 'You can leave your traps there,' he said, gesturing to his bag and coat. 'They'll be quite safe. Pirates we may be, but we don't thieve from our shipmates.' He reconsidered. 'Not usual, anyhow.'

Flynt did as he was told.

'And them barkers and that there sword stick, you can lose them too, if you please.'

Flynt was not willing to enter that room without weapons. 'And if I don't please?'

Hands grinned. 'Then we shall have us a disagreement and we can resume our little dance from earlier, only this time I shall not be so lenient.' He reached out, palms up, awaiting delivery. 'Though the doing of it would pain me something extreme, I won't hesitate.'

Flynt might have argued the point further but a deep voice, sounding as if it was carved out of the wood that surrounded them, boomed from beyond the cabin door. 'Bring those armaments to me, Hesikia, I would examine them for myself.'

The quartermaster crooked his upturned fingers. 'You heard the cap'n. I wouldn't be keeping him waiting if I was you.' He lowered his voice. 'We took us a cargo of spirits yesterday and he can be terrible intemperate when he is on the brandy.'

Another roar from the cabin. 'Mr Hands, what is the infernal delay?'

Hands' voice grew to a plea. 'Don't test him, landsman, nor less me. Give them barkers over and the stick too and you will have my word that they will be returned to you.'

The word of a pirate, as Captain Johnson had earlier stated, was not built upon a firm foundation. Nevertheless, there was something in this man that promoted trust. Flynt decided to acquiesce and received a grateful nod from Hands in return, before motioning for him to enter the cabin.

It wasn't a large room, though it was more spacious than Captain Johnson's quarters aboard the *Sprite*. The far wall was dominated by a wide window made up of small panes of glass revealing the open blue sea beyond. A broad rectangle of sunlight slanted into the cabin, illuminating a heavy desk on which rested a spyglass, a bottle of brandy and an open book, lying face down, but in the dim light Flynt could not make out the title. Beneath the book was a large parchment of yellowing paper, curling at the edges, and to one side was a goose feather quill resting in an inkwell. A single shelf of further volumes ran along the wall to the left of the desk, but again he couldn't discern their titles. A chair sat before him, a simple oak construction, but the one behind the desk was more ornate. It was of darker wood, the upright carved into a form of floral design, the arms heavy and similarly carved. A wide hat hung on one upright, a scabbard holding a sabre dangling from a broad belt beneath it, a double bandolier sporting three brace of pistols on the other.

But it was the man in that chair who drew the eye. Even though he was seated, and with the light at his back not much more than a silhouette, Flynt knew that he was tall of stature with a thick, powerful body. Without a word, he rested a straight-sided pewter tankard with a lid upon the desk and held out both hands to his quartermaster, who handed over Flynt's weapons.

'This here be Joseph Blake, Cap'n,' said Hands. 'He was soldier once, he says.'

The captain leaned forward then, the light falling on his face. Blackbeard, they called him, and in his case the cognomen was no lie. His facial hair was bushy and long, stretching down to his chest and, as far as Flynt could tell, rising almost to below his eyes, which were topped by thick eyebrows in which creatures could most comfortably set up home. His thick hair was unkempt, in parts curled into tails, some of which were tucked over his ears.

He turned Tact and Diplomacy over in his hands, making a close study of their workmanship. 'These are not the weapons of a common soldier.' He looked up at Flynt through the frame of dark hair. His eyes were bright and shrewd. 'You were an officer?'

There was a trace of the west country in his voice, but it was leavened with considerable learning. Captain Thatch was an educated man but he had not lost his roots.

'The highest I rose was Serjeant,' Flynt replied, 'and that not for very long.'

'Why not?'

'I didn't take to being ordered about, nor less issuing orders.'

'You don't like following orders?'

'Not when they are delivered by men who owed their rank to their family, or their fortune.'

The glance between Thatch and his quartermaster was difficult to translate. It might have been suspicion, it might have been one of approval, Flynt couldn't tell.

'You were infantry?'

'Aye.'

'And you served where?'

'Flanders.'

'Under Marlborough?'

'Aye.'

'You saw action?'

'Ramillies, Oudenarde, Malplaquet. A few other skirmishes.'

'Malplaquet was bloody, I'm told.'

'The bloodiest. That was when I decided that the life of a soldier was not for me, so I left as soon as I was able.'

'You deserted?'

'I reclaimed my liberty.'

Thatch grunted then resumed his examination of the pistols. 'These are custom-made pieces. Fine balance.' He held one out, sighted along the barrel. 'Did you steal them?'

'No, they were made for me.'

'For what reason?'

'I required weapons on which I could depend.'

'Why?'

'My occupation demanded it.'

Thatch gave him another long look. 'Highwayman, I'll be bound.'

'Among other things.'

'And now you wish to be pirate.'

'The opportunity arose, I took it.'

The captain laid the pistols down and leaned back in the high-backed chair. 'Why should I accept you?'

'You need men.'

'I need seamen.'

'You need men who can fight. I've proved that is something I can do.'

'So Mr Hands here has informed me. I'll admit I am impressed, for there are not many who can best him.'

'I believe we were evenly matched.'

'Even that is rare.' Thatch now turned his attention to the silver cane. He twisted the handle and drew the blade free. 'This is decided sneaky.'

'Sometimes the element of surprise is required.'

That provoked a hoarse chuckle. 'Aye, that is true.' He slid the blade back, locked the handle and laid the cane beside the pistols. 'What know you of our life?'

'I know that you live it free and easy. That's enough for me.'

'Free and easy, to be sure, but only to a point. We have chain of command, Captain – that is I – to Quartermaster, that is Mr Hands, and then to the crew. When an order is given in action, it will be obeyed with no back talk.'

'I understood life aboard a pirate vessel was more egalitarian.'

Thatch's prodigious eyebrows rose sharply. Flynt could have sworn he heard them rustle. 'Egalitarian? Now how come you by such a fine word?'

Flynt shrugged. 'I've read a book or two.'

Amusement crept into Thatch's dark eyes. 'You've read a book or two,' he said, thoughtfully. 'We have some Scotsmen on the crew and I'll wager none of them have read a book or two.'

'Not all Scotsmen are the same, Captain, just as not all pirates, or their captains, are the same.'

That generated another laugh, this time more open. 'Aye, isn't that the truth, eh, Mr Hands? Some of our fellows aren't worth spit. But you are correct, life in the brotherhood is somewhat more *egalitarian* than life on a merchantman or in the regular navy. We have our articles, which I will get to presently, and we have our common council, when the crew gathers to make decisions, even to depose their captain, if they have a reason and are minded to it, though it would take a brave man to propose that in my case, am I right, Mr Hands?'

'The men is most satisfied with your captaincy – they has no cause or complaint for plotting such a course.'

Thatch leaned forward. 'Votes to change leadership can be cast and decisions taken but not when we are in the process of taking a prize,

Mr Blake. For when action is underway it is my word, relayed through Mr Hands here, that is law and there will be no questioning of it by the crew. I am no idiot. I hold my office not through family or through purchase but through deed and deliberation.'

'I understand that, Captain.'

'See that you do, for I will brook no dissent at such times. I don't like murmurings below decks and comments passed from the side of the mouth. Once the prize has been taken and we are safe away, that is the time for questioning. But when we are in action, if I point my finger...' He raised his right hand, his finger extended, his thumb cocked, as though it were a pistol. 'And declare that a man must die, then that man dies and there is no questioning of it. Is that understood, former Serjeant Blake of the infantry?'

'You make yourself most clear, Captain.'

Thatch picked up his tankard again as his dark eyes gazed at Flynt. His study was piercing and in it Flynt saw a fierce intelligence. 'You have proved you can fight, but can you kill?'

'I have done so many times,' Flynt said. 'I don't enjoy it.'

'Killing is not to be done for the enjoyment but for the necessity of it. I know not what you may have heard of me but I try to avoid it wherever possible, but when it needs to be done, I will do it without a second's thought. But mark this, nobody dies unless I order it. We are a brotherhood, egalitarian as you say, and so you will find aboard this ship men of all nations, faiths, beliefs, colours. We make no distinction and there is no disagreement over how, or if, they worship a god. Speaking personally, I have no such beliefs. What of you?'

'None at all.'

A nod of satisfaction at that. 'We allow no women on board – I understand Mr Hands has already made that point. Liquor is also prohibited, for it fires the blood at all the wrong times.'

Thatch spotted Flynt eyeing the tankard and his teeth showed through his beard.

'The captain is granted some dispensations, Mr Blake, and I take full advantage of them. As you will learn, thievery from a crewmate is unacceptable. Fighting will not be countenanced. If there is a disagreement, then it will be brought to Mr Hands here and he will pass judgement. If it is serious then it will be put before the crew common council and the combatants can wait until we go ashore and settle their differences on the beach – fists, swords, pistols, we don't much care which but no man spills a crewmate's blood on these boards. If one man should kill another then that man will himself be killed. If I or the crew feel merciful then it will be a ball to the head but there are other, less compassionate, punishments. What do you know of keelhauling?'

'Nothing, but in truth I don't like the sound of it.'

'And well you may. The condemned man is bound and thrown over the side. He might drown, he might be lucky and break his neck or back when he hits the water. If not, he will be dragged under the ship, his body scraping against the barnacles. They are most unforgiving, those creatures. Hard little bastards, ideal sailing mates for men like us. The flesh is scraped off, the body tormented, and if he survives even that, then he will die soon after. I've only seen it once in my life and I pray I never see it again, eh, Mr Hands?'

The quartermaster agreed. 'It ain't pretty, Cap'n, and that be the truth of it.'

'It makes a flogging seem like mere discomfort, though most of these men are well acquainted with the lash, one way or another, so it holds little terror for them. Were you flogged in the army, Blake?'

Flynt shook his head.

'So you were never recaptured after you deserted?'

'Reclaimed my liberty,' Flynt corrected.

'Fancy words, but deserted you did, and though I can understand why, mark this, I will not accept it here. Deserters will have their ears and nose cut off and be cast away on a deserted island. There are many such rocks in these seas. Egalitarian, aye, but it's a disciplined egalitarian society, for without rules, with men such as these on board, then we would have anarchy.'

Thatch lifted the book, set it aside and pushed the parchment across the desk. 'These are the articles you will sign. You agree to the rules of the company and will honour them. In return you receive one share of whatever prizes we take while you remain on board. For your information, and to show that everything is free and clear, Mr Hands and I each receive one and a half shares apiece, the bosun and gunners and carpenter each a share and a quarter. There will be no dispute over the division of the shares when we reach port, it will all be done fair and proper. Any dispute will be referred to the common council for ruling.' Thatch dipped the quill in the well and held the feather towards Flynt.

'Your knowledge of the word egalitarian tells me you have your letters, so append your signature to this document and go on our account.'

Flynt took the quill and leaned over the desk, reading the document. 'And if I choose not to sign, knowing now all that I know of this life?'

'It's a long swim after the *Sprite*,' said Hesikia Hands. 'Even longer if we put out your eyes first.'

Flynt saw the sense of that. 'In that case, I'll sign, and right gladly.'

He scratched the name Joseph Blake at the end of the list of crew members, grateful that he had chosen to adopt it. If there were any repercussions for joining the pirate crew, irrespective of whether he ever had a share of spoils, at least it would not be Jonas Flynt they hunted.

Thatch pulled the articles back across the desk and examined the signature. 'A fine hand, Joseph Blake, a fine hand, indeed. It's uncommon in this life to find a man who not only can read but who ciphers most legibly.'

'My father said a man who had his letters was a man who would make his mark on the world.'

It wasn't Gideon who said that, for he was not the one who taught him to read and write. Flynt never knew his mother, for she died soon after he was born, and Gideon was often away at sea, leaving him in the care of an aunt who had instilled in him a love of literature, of Greek myths and Roman heroes, of philosophers and generals. She thought such pursuits would help tame the wild side of him, but it only gave air to the flame of adventure that eventually led him away from Edinburgh.

Thatch nodded. 'Your father was a wise man, is he still with us?'

Flynt nodded. 'In Edinburgh.'

A lie, of course, but if Toby Hawke traded in any way with these men then it was best to avoid the truth.

'We shall have further discussion, Blake, on books and such, for I am most starved of such discourse,' Thatch said. 'Mr Hands here is a fine sailor and a superb quartermaster and he can read and cipher well enough, but those hands were built to wield cutlass and pistol, not to hold a book, am I right, Mr Hands?'

Hesikia smiled in agreement. 'Reading is just a waste of good drinking and wenching time, Cap'n.'

Thatch laughed. 'There is time for all, Mr Hands, as I can most readily testify.' He raised the tankard and swallowed deeply from it before he regarded the parchment once more and the humour vanished.

'You have signed these here articles, Mr Blake, and that makes you bound and dedicated to this ship, this crew and its captain. You've shown some skill with the fighting arts, now we shall put to the test your resolve.'

Something in the manner of delivery of those words gave Flynt pause. 'In what way?'

'You shall see. Mr Hands, has common council been convened?'

'The men await you on deck, Cap'n.'

Flynt sensed something hanging in the air. 'Won't you tell me what is afoot?'

Thatch stood, retrieved his hat from where it hung over the upright of his chair, strapped on the belt with the sabre, then plucked up the bandoliers. 'Justice is afoot, friend, justice…'

20

Thatch hauled the bandoliers over his head and fixed them across his chest as he led them onto the deck where the sunlight blinded Flynt for a moment or two after the gloom of the cabin. He found himself for the second time that day surrounded by a ship's crew, only this time there was no sense of the urgency, or fear, that he had experienced on the *Sprite*. This time the mood was sombre, with perhaps even a little anger, but there was something else, too. Grim determination.

Justice, the captain had said, and a test of Flynt's own resolve. Whatever lay ahead, he wasn't going to like it.

'The quarterdeck, if you please, landsman,' Hands muttered behind him. Flynt followed Thatch's tall figure to take up a position beside the large double steering wheel manned by a heavily muscled man from the Far East. As he stared over the rail at the faces below Flynt saw that Thatch's claim about his crew being from all races and colours was true. The faces angled towards them were a mix of white, though burnished by the sun, black, brown and, like the helmsman, from the east. In the centre of the throng, given some clearance of a three-foot radius, stood one man, his head bowed, his wrists tied in front of him by thick rope.

Justice.

Flynt's disquiet increased.

Captain Thatch rested both hands on the rail and surveyed his men. 'Lads,' he began, 'we has a solemn duty to perform this day, one that – and for this we be utmost thankful – we seldom has to undertake.' His accent had coarsened a little, though it was not nearly as rough as the quartermaster's. 'Life at sea be hard, the life of the brotherhood be hard, and we must depend on the man beside us to support us. We must trust him most implicit, do we not?'

There was a murmur of assent from the men.

'But when that trust be undermined, what must we do then, eh, Mr Hands?'

The quartermaster, standing beside Flynt, spoke up. 'We must act, Cap'n.'

'We must act and do so swiftly and with resolution.'

We shall put your resolve to the test, Thatch had said.

Justice...

Flynt's eyes found the man standing alone in the centre of the assembly once again, an ugly suspicion forming in his mind.

Thatch pointed a dramatic finger at the man. 'This here man, John Sanders by name, carpenter's mate by trade, has been sailing under our colours for two years now. He was with us when we served with Captain Hornigold. He was with us when we took this here fine vessel from the Frenchie. He was friend to many of you. He served well, did he not?'

Responses of 'aye' and 'a good sailor', louder than before. Heads nodded. Some swivelled to stare at the man called Sanders, who still maintained his downward glance.

'But then he committed a sin, a cardinal sin aboard ship,' Thatch continued. 'He stole from a shipmate. A trifle it were, a bauble only. A locket, a keepsake from a past the shipmate had left behind, but one that he wished to keep close, for he had tender feelings attached. So a gimcrack to us, but important to you, was it not, Dicken Trelawny?'

This last was addressed to a small, powerfully built man in the forefront of the congress, who stepped forward. A deep scar ran from forehead to chin, the eye it crossed milky white. 'It were my mother's,' he said, turning to address his crewmates. 'It were the only thing she had of any value, and even that weren't much. It were taken from my traps two nights since.'

'Aye, and found in the possession of John Sanders yonder,' accused Hands.

'He had often admired it,' Trelawny continued, his accent steeped in the salt of the Cornish coast. 'Had once offered to purchase it from me, for he said he had a girl on New Providence who would look with favour upon him if he presented it to her.'

Thatch asked, 'But you refused to sell, Dicken, is that right?'

'That be the truth of it, Cap'n. I told him straight, I said, I'll never sell it, John Sanders, I said, and certainly not for it to be given to some dockside doxy in Nassau town. It were my dear mother's, I said, and it be the only thing remaining to remind me of her, for she lies a world away in the cold earth.'

The men shifted their feet and more accusing glances were directed towards Sanders, who remained as still as a statue.

'You all signed articles,' Thatch's voice boomed. 'You all agreed to them. We have rules aboard ship, you all know what they are. And what be one of the principal rules, can you tell me?'

The men knew immediately that to which he referred, for the majority intoned, 'We steal from others, not from brothers.'

'Aye, that be the one, lads. We steal from others, not from brothers. There are many out there who would wish to take from us what we have. The governors, the navy, even fellow raiders on occasion, so if we cannot trust the man serving at our side, then we are doomed. We need to know that each man who serves with us has our backs, that we can depend on him, that we can trust him most implicit.' He paused, once again placing both hands on the rail and leaning over just a little. 'We steal from others, not from brothers. Break that trust and there is nothing left. John Sanders broke that trust and, if this here council wills it, he will be punished, by regulation and by ship's law. What say you?'

The reply of 'aye' was loud and firm.

Thatch held up a hand to silence them. 'John Sanders, you heard the will of the council, of your shipmates. What do you have to say in your defence?'

Most heads turned to face the man again and silence descended. At first Sanders didn't move but eventually, he looked up and tears streaked his sun-darkened cheeks. It was a shock to see that he was just a boy, little more than Jack's age, or Barbecue's.

'It were a madness what took ahold of me,' he said, his words trembling, his voice hailing from the back alleys of London. 'I doesn't know why I did it. It were there, in Dicken's traps, one moment and the next it were in my own hand, like something magical. Dicken speaks true, for I had admired it previous and thought it would be a fitting tribute to my girl.' His voice strengthened. 'But she ain't no dockside doxy, on that I has to correct Dicken most emphatic. She works the tables in the Mermaid to be sure, but I state with fervour that she's a good girl. I has proposed marriage to her and she has accepted and I didn't have no token to give her, so I sees the locket and... well...'

He lapsed into silence. All eyes were on him. Above them, the wind sighed through the rigging and puffed the sails. The boards creaked.

Finally, Thatch spoke again, his voice considerably softer than before. 'Theft be theft, boy. And there be only one punishment. Before I pass sentence, is there any man here who wishes to speak on the boy's behalf, to plead for mercy?'

He waited, but nobody came forward. Flynt felt he had to say something but the quartermaster laid his hand on his arm and shook his head. 'Say nothing, landsman. This will be what it must and nothing you say will change it. You ain't got the right yet.'

Thatch took a breath. 'John Sanders, carpenter's mate, you has confessed freely to this here congregation to be guilty of theft from a shipmate, a crime for which there is only one punishment. As captain, and bound by articles, it is my duty to pronounce sentence. You know what it is, boy.' He paused, removed his hat and held it before him in two clasped hands, as if he was praying. 'The sentence is death, to be carried out immediate. In respect to your previous good service and the fact that your time aboard ship has been unblemished, there won't be no keelhauling, it will be swift and merciful. And if you have a god, boy, then I hope he shows you favour.'

Sanders had known the outcome but he still flinched when he heard the sentence. He sank to his knees, his head bowed.

Thatch half turned towards Flynt. 'Mr Blake, please step forward.'

Flynt did as he was told, dreading what was coming.

'Mr Blake here was soldier and thief, and has now signed articles. It be my wish that he show his loyalty to us by carrying out this sentence.' He gestured towards the young man isolated among his former friends. 'Mr Blake, if you please.'

Flynt stared at the captain's hair-dark face, at the eyes watching him closely from under thick brows. This was a test, Flynt recognised that, but one where a young man would lose his life. If he refused, then he may never reach Nassau. If he carried out the sentence of death, then he would have taken the life of a man with whom he had no quarrel. With one squeeze of a trigger, he would take away this young man's – this boy's – past, his future, his hopes and dreams.

Thatch sensed his thoughts. 'Shoot straight, Mr Blake,' he said quietly. 'Let me see your commitment to this company. Show me your resolve.'

Without a word, Flynt stepped down onto the upper deck. The men before him parted to let him pass. Nobody spoke. Some avoided his gaze. Others were sympathetic towards his unease. A few were openly hostile. They were hard men, well used to the rigours of their trade, but they were not all heartless. They had served with this young man, many would no doubt like him, and this stranger was about to kill him. They would know that he had little choice, but all the same it should be one of them who put an end to him.

He heard the boy's sobs as he grew nearer. When he saw Flynt's boots, he raised his head, his eyes awash, liquid streaming from his nose. He blinked towards Flynt's face, then saw the pistols in his belt and his breath quickened.

'It was madness, is all,' he pleaded. 'I wouldn't have kept it, I would have give it back to old Dicken, honest I would. I just lost my head momentary...'

Flynt wanted to tell him that everything would be well. But he couldn't.

Thatch's voice came from behind him. 'Carry out the sentence, Mr Blake.'

Then another voice rose. 'This ain't right, Cap'n...'

Rufus, the man who had discovered Cassie and young Jonas on board the *Sprite*, shouldered his way through the men.

'Regulations is regulations, Rufus,' Hands said. 'The lad admitted to the theft and he must pay the price.'

'Aye, perhaps, but the punishment and the crime don't deserve each other. We all knows about that there locket of Dicken's. It ain't worth a fart, not in monetaries. A fine be sufficient, and I ain't the only one to say this.'

'The cap'n has ruled, Rufus, so stand down...'

'This be common council, and any voice can be heard at common council, ain't that right, boys?'

He looked at the men around him, one or two nodding in agreement, others remaining silent, studiously so.

Thatch leaned over the rail. 'Are you petitioning for clemency, Rufus Turner?'

'That I am.'

Thatch mulled this over, then turned towards the steps leading to the deck. 'Do we have a seconder?'

There was a note of warning in his voice that did not go unnoticed by the men, for nobody stepped forward. The throng parted to allow the captain to approach. 'It seems you are the only one who believes mercy is warranted, Rufus.'

Rufus looked around him, the tension of his lips suggesting that he knew there were others who believed as he but who had declined to come forward. 'I ain't the only one. The others be too scared to speak up for themselves.'

Thatch held out both arms and swivelled in a half-turn to the men around him. 'Scared? Of what? This is common council – any man can speak his mind here.'

Flynt began to wonder how egalitarian life under Thatch truly was.

'Aye,' said Rufus, now showing signs of doubt. 'And that be what I'm doing, Cap'n, not meaning no disrespect. But I say again that this ain't right.'

'The council has spoken,' Thatch said, 'and I have ruled.'

Rufus pointed towards Flynt. 'Then at least let it be carried out by one of John's shipmates, and not some landsman stranger.'

'This be the way I wants it. Unless you wants to try to depose me and countermand my order.'

Rufus seemed to find that idea appealing, if only briefly.

Thatch leaned closer to him. 'But for that you would need a seconder, and I don't hear no other voice speaking up, do you?'

Rufus again searched the men nearest him, but they all avoided his gaze. The man's mouth worked furiously as he chewed on anger and trepidation.

'Of course,' Thatch said, taking a step back, his hand falling on the hilt of his sword, 'you could try a more direct challenge to my authority. I'd be most pleased to accommodate you.'

For a moment, Rufus seemed to consider a challenge but then thought better of it. He kept his hands well away from his weapons. 'Not me, I've no desire to do such.'

'Then may we proceed with due execution of the sentence?'

Rufus was unwilling to back down fully. 'I just wants my voice heard, saying that I don't believe this to be right and proper.'

Thatch turned his back on him, but Flynt could tell he was tensed for an attack. 'Duly noted. Mr Blake, you may proceed.'

The only visible sign of aggression on Rufus's part was a face now dark with anger and perhaps shame; then, with a final glance towards Flynt and the boy, he pushed his way through his shipmates to vanish into their ranks.

Sanders watched Thatch as he sauntered back to the quarterdeck, then he looked at the faces around him, as if seeing them for the first time. If his hopes had risen during the exchange, he showed no sign, but during the interruption he had apparently come to understand that he could do nothing to prevent what was coming. His tears had halted, he had wiped his nose with his bound hands.

'Mary Bowers,' he said, looking back at Flynt. 'My girl. Mary Bowers, in the Mermaid tavern. I'd be grateful if you would ensure my traps gets to her. There ain't much to show for my life but I wants her to have what there is, and if you would accommodate that, sir, I would be most grateful.'

Swallowing back his revulsion for what he had to do, Flynt nodded. 'Tell me about her, John.'

The fear in the young man's eyes eased a touch as he thought about Mary Bowers. 'I loves her, I truly does...'

Flynt half turned from him, his hand inching towards a pistol. Sanders must have seen the movement and knew what lay ahead but he made no comment, nor any sign of understanding, apart from allowing his eyes to return to the deck.

'I never thought I'd say that to a girl,' he said, 'but there you has it. But tell her I did, free and sincere...'

Flynt eased Tact from his belt, saying gently, 'Picture her in your mind, lad.'

A soft smile played with Sanders' lips. 'She is a sweet girl...'

Flynt brought the pistol to bear.

'...flaxen-haired, and the kindest disposition you could ever—'

John Sanders died with the image in his mind of Mary Bowers, the flaxen-haired girl who waited for him.

21

Flynt was invited to dine with the captain and Quartermaster Hands that evening in Thatch's cabin. The massive desk had been converted into a dining table, resulting in Flynt having to sit with his legs in a sideways position thanks to the thick side panels. He sensed it was a great privilege for a neophyte crew member to be so invited, but he didn't feel much honoured. All he could think of was John Sanders' face as he turned away and spoke of his girl, knowing what was coming. Once again, Jonas Flynt had done what was necessary, and once again he had lost a bit of himself in the doing of it.

Thatch drank steadily but displayed no sign of its effects. Under that thick mat of hair, his skin might have been flushed but his demeanour betrayed no evidence of being the worse for the liquor, and his speech remained unslurred. Flynt accepted a goblet of wine, for grace's sake, but sipped it sparingly throughout the meal, at which he had picked. In truth, he had little stomach for either food or drink after the afternoon's events. He thought perhaps his comportment would go unnoticed. He was wrong.

'The pork is not to your taste, Mr Blake?' Thatch asked.

'The pork is excellent,' Flynt said.

'And the wine? You do not like it?'

'The wine is superb.'

'It should be, for the French merchant from whom we wrested this ship put up quite a fight, didn't he, Mr Hands?'

'There's nothing a Frenchie defends more fervent than his wine, Cap'n,' said Hands.

'They do like their wine, that is true. But I managed to point out the error of his ways.' Thatch was silent as he studied Flynt once again. 'You are troubled by what I had you do this day, perhaps, Mr Blake?'

The man had opened the door, so Flynt stepped through, not caring if he caused offence. 'Was killing the boy necessary?'

Thatch was untroubled by the question. 'Necessary? Aye, it was.'

'For what reason?'

'I told you before. Discipline. If we don't have discipline then we have nothing. If I was to let that crime, petty though it was, go unpunished then it would lead the men to believe that such transgressions are permitted. They are all thieves, Mr Blake, never forget that, but Mr Hands and I have to marry them together to make a crew and only discipline can do that.'

'With respect, having me carry out that sentence was not solely for the sake of discipline, you said as much yourself.'

'Aye, I wanted to test your mettle, that's the truth of it, and you passed. You did what had to be done.'

As I always do, Flynt thought bitterly.

'However, if it means anything, had you not then someone else would have, even if I had to do it myself. Young John knew he had wronged the entire crew when he took that cheap tin locket of Dicken's. He broke faith with the articles against which he had made his mark. I liked young John Sanders, and what happened today was the most merciful thing I could do under the circumstances.'

Seeing that Flynt remained unconvinced, Thatch pushed his own goblet and plate aside to lean over the tabletop towards him.

'You were soldier. You saw punishments meted out for the smallest infraction of regulations, correct?'

'Yes.'

'That's because you and your fellow soldiers were in service to king, to country, and as such had to be taught that to disobey, to break regulations, has repercussions.'

Flynt had witnessed brutality meted out for little reason, and had never agreed with it. 'These men are not in service to king and country.'

'But they are. I'm their king, and while at sea the *Revenge* is their country. This mass of wood and nails and canvas is a floating country with its own laws. And I will uphold those laws, even if it means having to spill the blood of a man I likes.'

He spoke casually, conversationally, but Flynt heard the warning in the words. If he stepped out of line he could expect the same treatment as John Sanders.

Thatch sat back again. 'Sometimes I have to kill one of them just to remind them who I am, is that not so, Mr Hands?'

'All done legal and proper according to ship's rules, Cap'n.'

Thatch raised his goblet. 'Aye, as quartermaster you make sure of that, do you not?'

'The men expect me to represent their wishes before you, just as you expect me to convey your wishes to them.'

'And what do they say about today's events?'

Hands chose his words carefully. 'Most are in accord.'

'Most?'

'There be some who don't like it.'

'Some, like Rufus Turner?'

'I can handle Rufus, Cap'n, don't you be fretting.'

Thatch fell silent. A strange atmosphere filled the room and Hands began to fidget in his chair. The captain's demeanour had changed. He said nothing, did nothing, merely sat as he had before, one hairy hand wrapped around his goblet, his eyes fixed on his quartermaster. Finally, he said, 'Why don't you fetch Mr Turner to me?'

Hands raised a finger. 'Cap'n, leave him be, is my advice. Rufus is all bluster. He wanted to bring a wench aboard from the *Sprite*, I told you such, and that still grates upon him. And he liked young John, we all did...'

'So you say I was too harsh? You agree with Mr Blake here?'

Thatch's voice had lowered, almost to a growl, and his eyes had taken on a strangely luminous quality. The menacing atmosphere intensified. Flynt recalled Hands' earlier words, about the captain being terrible intemperate when imbibing.

'Articles is articles, Cap'n, you know I agrees with such, and John, well, he was recognisant that he had crossed the line. He knowed the penalty for thieving from a shipmate and he paid it.'

'But Rufus Turner still questions?'

'As I says, Rufus is all bluster and—'

'Bring him to me.'

'Cap'n, I does beseech you to leave this be. There ain't nobody listening to him...'

Thatch shot to his feet with such suddenness that Flynt automatically reached for a pistol. 'Damn it, if you will not bring him to me then I will go to him.'

He flew past the still-protesting quartermaster and threw open the cabin door. Hands swore softly and followed behind. Flynt was momentarily perplexed by the rapid change in the wind, but he quickly recovered and went after them. Thatch stood in the middle of the deck, the moon high and bright in the velvet sky, casting his eyes around him, searching for one man.

Not seeing him, he bellowed, 'Turner! Rufus Turner! Where are you, man?'

Hands was at his captain's side now, still trying to dissuade him from any further action, but Thatch pushed him away. 'Turner, God damn your soul, show yourself.'

The crewmen on duty stopped working to stare at their captain, some exchanging querying looks.

'Cap'n, don't be doing this…' Hands pleaded. 'It ain't the right and proper way to…'

'To hell with right and proper. I'll have this settled this night, for I am weary of Turner's continual bitching. You know he has eyes on my position and I've ignored it too long.'

'But…'

'No, Mr Hands, do not oppose me on this, damn you. I shall have this out with him and…'

He stopped, his eyes fixing on the companionway leading from the crew's quarters below. Turner's head and shoulders had emerged but he paused in his ascent.

'I see you, Rufus Turner,' Thatch said. 'Pull yourself from that there hatch and we shall converse.'

Rufus frowned. 'What about, Cap'n?'

'What about?' Thatch laughed, but there was little levity in the sound. He looked to his quartermaster. 'What about, he asks.' Another laugh, as he regarded Turner, who had now climbed fully onto deck. 'You know full well what it's about.'

'No, Cap'n, I…'

'By God, I've never seen such innocence in a fellow before. You might be an altar boy with such a face, if you weren't such a back-stabbing, bilge-sucking son of a whore.'

Even in the moonlight, Rufus's expression considerably darkened. 'You ain't got no call to address me in such a manner, Cap'n.'

'I don't, eh? That be so? Well, Mr Turner, I believe I do have call to address you in such a manner, what say you to that?'

Rufus looked to Hesikia. 'Quartermaster, are you going to let him speak to a crew member in that way? He's addled with liquor, and that right there is a breach of his own articles.'

Hesikia was clearly uneasy but he backed his captain. 'Cap'n always has certain leeway in such matters, Rufus, you knows that.'

'Leeway, is it? Rules for some but not for others, is what that be, ain't that right, boys?'

The deck was filled now with much of the crew and Rufus looked for support, but as before nobody came forward. Thatch lowered his head. 'You're on your own, it appears.'

Flynt disagreed. He had witnessed two men nodding surreptitiously at Rufus's words.

Rufus shrugged. 'There's those that agree with me; I know it to be true.'

Thatch's fingers edged towards the lower brace of pistols on his bandolier. 'Do you challenge my authority?'

Rufus didn't reply, but his previous assured manner faltered.

The captain took a step forward. 'Speak up, man. You've been murmuring a-plenty below decks, I'll be bound, so why not give voice now? Do you challenge my authority?'

Rufus backed away a couple of steps. 'This ain't the time to be taking such action. Common council is the place.'

Thatch waved one hand. 'Look around you, your crewmates are all here, most of them, anyway. This be common council sufficient for this purpose. So, I ask again, do you challenge me for captaincy? Do you think you can do a better job of commanding this here vessel? Find fatter prizes?'

The quartermaster stepped between his captain and Turner. 'This be most irregular, Cap'n. Why not sleep on the issue and return to it fresh at sunrise?'

Thatch pushed him out of the way. 'We settle it right here and right now. This man has been a constant grumbler since the day he joined the crew, arguing against one decision or another. Well, I won't be having it no more. So speak up, damn you. Do you challenge me?'

Rufus was nervous. His eyes met with the two men and one of them gave him a little nod, his hand resting on the pistol in his belt. Rufus's own hand twitched towards his weapon. Thatch smiled. 'You want to tug on that there barker?'

Rufus seemed frozen.

'Go ahead, man,' Thatch taunted. 'I know you want to. I know you think you're better than I. I know you want your shipmates to see that you're the man who bested Edward Thatch, old Blackbeard, who you think ain't fit and proper to command this here vessel.'

Rufus shook his head, but his hand didn't leave the butt of his pistol. Behind Thatch's back, the two crewmen almost casually drew their pistols and let them dangle loosely at their sides. Nobody but Flynt seemed to notice.

'Do it, man,' the captain urged. 'Pull that barker. Let's see if you have the guts to back up your words...'

Hands stepped in front of Thatch, his back to Turner. 'Stop this, Cap'n, this ain't the proper way...'

'Step aside, Hesikia, this has been brewing for a considerable time.'

Rufus seized his chance to draw his weapon. Thatch snatched his own from his bandolier but Hands gripped his wrist. 'No, Cap'n...'

'Damn you,' the captain snarled.

Rufus fired, too hastily, his nerves getting the better of him, and the ball caught the quartermaster in the back of the leg. Blood ballooned from his knee and he went down with an agonised scream.

The two crew members each raised their weapons and aimed at Thatch's back. Flynt, having surreptitiously drawn Tact and Diplomacy, without hesitation shot them both. Thatch whirled into a crouch, his pistol ready to fire, saw the men sliding to the deck and so gave Flynt a brief nod in thanks. He straightened, stepped over his wounded quartermaster and drew a bead on Rufus, who stilled with his own second pistol only half raised.

'Never fire in haste, man,' Thatch said, advancing on him. 'Deliberation is what's needed when you really want to kill a man.'

Rufus looked around him for a means of escape but saw nothing.

'You have a second barker there, why don't you use it?' Thatch suggested, still moving closer, the pistol in his outstretched arm held steady. 'You never know, I might miss. I've had a surfeit of liquor, as you pointed out, so my aim may not be true. That would give you those precious moments to aim and fire, slow and easy. You can't miss now, not at this distance. Why, you could reach out now and place that barrel right against my forehead.'

Rufus cast his eyes about for help, but again, none was forthcoming.

'They won't help you. None of them. They saw what happened. You pulled first, I made to defend myself and the quartermaster interceded. Foolhardy it were, but he's a good man, is Mr Hands, a respected man, by me, by the crew, but you put a ball in him, blew out his knee by the looks of it. That's a death penalty, that is, and not an easy one, like young Sanders. It's the kiss of the barnacle for you. So perhaps you'd be best advised to take your chances with that there pistol, eh? What say you?'

Rufus looked beyond him to where the quartermaster writhed on the deck, attended by two crew members.

'So what's it to be, Mr Turner? Will you show yourself to be a deliberate man, or will you be the craven little sea snail I know you to be? Would you rather let ship's justice have you or throw the dice here and now?'

Rufus hesitated, then allowed his arm to fall. Thatch stared at him for a few moments, then lowered his arm. 'I thought so. All mouth below decks, no guts above them.'

He turned away slowly and Flynt saw what was about to happen before either of them moved again. Rufus's hand snapped back up, the pistol cocked but trembling in his grip. Thatch halted, but didn't turn. 'Why not pull trigger, eh?' He held both arms out from his sides. 'I'm an easy target, am I not? You have my back, isn't that what scum like you prefer? Take the shot.'

The silence that descended over the ship was like a thick blanket. All sounds vanished. The creak of the timbers, the hiss of the foam as the bow sliced through the surface, the billow of the sails as they caught, and momentarily lost, then caught again the wind.

'I didn't wish this,' Rufus said, casting his eyes around him, ever in search of support. Or perhaps sanctuary.

'But here you have it, so take the shot, man, and be done with it.'

Thatch was calm, sure of himself, nevertheless Flynt unlocked his sword and edged forward. If Rufus showed the slightest inclination to discharge that firearm, Flynt would see such an intention in his eyes before it reached his finger, but in order to do anything about it, he needed to be closer.

Thatch saw him move and gave him a brief shake of the head, as if he knew something nobody else did. He was smiling. Frowning, Flynt came to a halt. Some of the other crew members began to produce weapons.

'Easy, lads,' Thatch said, calmly. 'Keep them barkers where they be. I'll deal with this.'

Rufus showed no signs of pulling the trigger, yet he remained stationary, his pistol extended. Thatch came to the same realisation, turned and strode back to him. He wrenched the pistol from his grasp with one hand before delivering an open-handed slap across his face with the other, then a second blow to the man's other cheek with the back of his hand. The blows cracked in the still air.

'Gutless, is what you are, Rufus Turner,' he said. 'Gutless and useless and I'm ashamed to have you on my crew. Get yourself gone from my sight, for you are an insult to my eyes. You'll be dealt with later.'

His face vibrating with what Flynt assumed was shame and fury, Rufus turned and slumped back to the companionway. Thatch moved to his quartermaster with no further backwards glance. Flynt, however, kept his eyes on the hatch, in case Turner decided his pride required him to return and finish the job, but when he didn't appear, he joined Thatch and the crew members around the recumbent Hesikia Hands.

'He's passed out,' said one of the men.

Thatch's brow was furrowed as he examined the wounded leg. From what Flynt could see it was a bad one, with the knee almost completely obliterated. 'Get him to my cabin,' Thatch ordered, 'and have cook attend to him.'

Further assistance was requested and a total of four men carried the unconscious quartermaster aft. The bodies of the dead were hefted over the side without ceremony. Murmurs sweeping around the deck told Flynt that they would not be missed.

Thatch turned to Flynt. 'Thank you kindly for your service, Mr Blake.'

'It seemed like the thing to do, Captain. But should you allow Mr Turner his liberty? Should he not be taken into custody?'

Thatch looked back towards the hatchway through which Rufus had vanished. 'He won't be going nowhere. I was never in any real peril, not from those sons-of-bitches.'

'I'm not so certain, Captain. Those two men were within seconds of pulling the trigger, and while the man Turner was not so quick, he did seem exceeding agitated.'

'He's not a deliberate man. I know him, know his kind. All notion, no motion, not under his own volition. He'll take an order, and he'll complain most vociferous, but he won't take the final step.' He turned back to Flynt. 'As to those other two, that ain't how I will meet my end. I know for a fact how that will be.'

'How can you know such a thing?'

Thatch took Flynt's arm and led him towards the rail, where he was silent for a moment as he stared out into the darkness, one hand holding a stay leading aloft. 'It was down in Hispaniola, two years ago maybe, perhaps less, I found myself beached, temporary like. I met this woman, an old woman, African she were, too old to work the plantations, was of no use to her former owner who wouldn't spend the money on housing and feeding her, so she was set loose and made a living in the telling of fortunes in Santo Domingo. I was keeping out of the way of the Frenchies, you understand, so I hid in her little parlour. I meant her no harm and she knew that, so she welcomed me. She didn't have no love for the Frenchies either, not that much for the English, but she welcomed me. She was what they called an Obeah woman, you heard of it?'

Flynt shook his head.

'A religion, brought over from Africa, mixed a little with our own scripture. Anyways, she claimed to have the sight, knew the future, and she held my hand for a moment, stroking of my palm like a lover...' Thatch looked at the palm of his right hand. 'I can feel it even now, the heat of her touch travelling up my arm, to my chest, to my head. She told me I'd have my own command right soon, and she was right, for it wasn't long after that I was given me my first ship by Ben Hornigold. She said that I would become a force to be reckoned with, but I would die on the deck of a ship, and not one under my command. It would be daylight, she said, but I wouldn't know it and I would be enjoying victory but it would then be snatched away from me by the sword of someone I'd wronged.' He laughed. 'Now, that last part is a hefty long list of men, I'll be bound. Turner certainly fits that description but we were on my own deck, there's no sunlight, we be in open waters and that cur reached for no sword.'

Flynt recalled Belle St Clair telling him that she had consulted a cartomancer, what he'd disparagingly called a spey-wife. She was told that someone would come between them. He'd mocked the very idea of the future being in the cards, or in the palm of a hand, but then Cassie had returned to his life and had, indeed, come between them. He still retained considerable scepticism but he kept his silence on the matter. It was clear that Thatch believed it.

'I'm surprised you didn't kill Turner,' Flynt said. 'He could come back at you again.'

Thatch gave Flynt an enigmatic look and patted him in a brotherly fashion on the shoulder. 'Don't you be fretting about Rufus Turner, Mr Blake. He's shot his load and he knows it. There's a price to be paid and he'll pay it, of that you can be sure.'

Thatch left him and walked across the deck to the door leading to his cabin.

The following morning, Rufus Turner was nowhere to be seen. His traps remained below, but his bunk was empty and a search of the ship revealed no sign. A sailor did find traces of blood near the bow, and another on the rail nearby. It was surmised that somehow Rufus had injured himself and then fallen overboard. Nobody suggested coming about to find him.

Thatch was supervising two crew members who were scrubbing at the bloodstains, his dark features impassive, when he saw Flynt approach and said, 'It's a dangerous life on ship for the unwary, Mr Blake.'

Rufus may have fallen, or jumped overboard, it was possible, but Flynt knew a little of how blood fell and from what he saw being wiped away, this was not the result of a man having hurt himself and stumbling. The blood had sprayed in a jet across the rail and that suggested an artery had been slit, perhaps Rufus's throat. Flynt gave Thatch another glance. He'd said the man would pay for his actions, and it seemed he had. His words from the night before echoed in Flynt's mind.

Sometimes I have to kill one of them just to remind them who I am.

22

The *Revenge* was two days away from Nassau when the lookout high on the mainmast spotted the sail. He sang out its location and Captain Thatch, on the quarterdeck, called for his glass and swept the horizon for sight of it. He was very still for what seemed like an inordinately long time, studying the ship as it and the *Revenge* shortened the gap of sea between them. Finally, he slammed the glass shut and handed it to the nearest seaman, who waited for further orders.

'It's old Hornigold's ship out of Nassau – I'd know it at any distance. You know of him, perchance?'

'I've heard the name,' Flynt replied.

'Ben Hornigold, my former captain, gave me my first command, this here ship, which we took from the Frenchies.'

Flynt recalled Hesikia saying that the day they intercepted the *Sprite*.

'Aye, headed for Martinique from Africa it was, and loaded to the gunnels with precious cargo, gems and gold dust. That was a prize worth the effort of taking her, and he reached the reckoning that he had made his fortune, what with all the other prizes he took, and wished a quiet life, for he lost his taste for the raiding. It's a crying shame, for a fine master he was, fearless. Now he mostly sits in the fort above Nassau, him being a wanted man, and acts as governor, though without the nod of fat old George back in London.'

'It looks like he's left port now.'

'Aye, for a rendezvous with me, I'll be bound, for there's no other reason for him to leave the safety of the island.'

When Captain Benjamin Hornigold climbed the rope-and-wood Jacob's Ladder to the deck of the *Revenge*, having been rowed across from his own vessel, he did not appear to be in any way weak or weary. He was not tall but he was robust and his bearing erect. He dressed well, too. Whereas Thatch's black coat and breeches showed signs of wear and repair, Hornigold's clothes were of fine quality and well cared for. His hat sported a white feather and the sword at his side glittered with gold inlays.

There was another man with him, who barely gave Flynt a second glance as he ascended to the quarterdeck. He was older than Hornigold, not quite as well groomed. His thinning hair grey, his eyes sunken, his gaunt face carrying a scowl that Flynt suspected had become fixed thanks to overuse.

'Ned,' said Hornigold, clasping Thatch's hand.

'Ben,' said Thatch, and Flynt could tell there was genuine warmth between the two men. Thatch looked past him to the second man with considerably less pleasure. 'And seldom it is that I see you out here on the foam.'

'I've returned just recent from Scotland,' the man said, his voice carrying the same twang as had Rufus Turner's. 'A meeting with our friends.'

Flynt forced himself not to react. Scotland. A meeting. He made a closer study of the man.

Thatch seemed none too impressed. 'I don't trust those gentlemen, nor any gentleman of commerce. They sit on their fat backsides in their fancy homes in London and New York and issue orders to the likes of us, and do nothing else but rake in the profits.'

'You need those gentlemen of commerce, Captain Thatch. We all on New Providence need their influence...'

Gentlemen of commerce. Was that an oblique way of referring to the Fellowship? And could this man be Toby Hawke? Moncrieff had told him he'd met with Hawke in Edinburgh. This had to be he. Without making it apparent, Flynt sought the source of Dan Hawke's features in this older man, but apart from the same haughty cast to the eye, saw none.

Hornigold gave Flynt a glance, his face betraying a fear that Thatch was saying too much before a stranger, and cut in. 'And who is this, Ned?'

'Joseph Blake, who did me a fine service recent, when some of my crew decided they would relieve me of my command without proper recourse to common council.'

Hornigold's eyebrows shot up. 'Mutiny?'

Thatch laughed. 'To call it such would be to elevate its status. It were nothing but a pair of weak-minded lads who listened to the wrong man.'

'And where are they now?'

'Two be dead, and go on Mr Blake's tally.'

'And the one they listened to? In chains, I suspect, awaiting trial by the crew.'

Thatch smiled. 'He took his leave of the ship.'

Hornigold understood immediately. 'I thank you, Mr Blake, is it?' Flynt nodded. 'Ned here is a good man and a fine shipmate.'

Flynt inclined his head in appreciation. 'It seemed like the thing to do, Captain Hornigold. I don't like men who prefer to shoot in the back.'

The eyes of Hornigold's companion narrowed. 'You are Scotch?'

'Aye,' Flynt said. 'From Glasgow.'

The man was clearly suspicious. 'We should go below, for we have much to discuss.'

'My cabin is occupied at the present, I regret to say,' said Thatch. 'Hesikia was sorely wounded in the affray and my quarters have become infirmary.'

'I'm right disturbed to hear that. Hesikia is a good man.' Hornigold was truly sorry. 'He will recover, I trust?'

'A ball took his kneecap right off. The leg may have to be lost.'

'Putrefaction?'

Thatch nodded. His voice grim, he agreed, 'Aye.'

The other man grew impatient with the course of the conversation. 'We have matters to discuss.'

'Then proceed – nobody stops you, Hawke.'

So it was he. Flynt wanted to step closer but didn't. Not allowing his hand to drift towards the butt of a pistol was more difficult.

Hawke shot Flynt a look. 'I have matters that are not for the ears of strangers.'

'Mr Blake here is no stranger. Didn't you hear? He stopped two craven dogs from taking my life. Furthermore, he has signed articles aboard my ship, so he is now fully on account.'

'Nonetheless...' Hawke began.

'Nonetheless, Toby Hawke, this is my ship, and Mr Blake remains and that be the end of it.'

The underlying edge between Thatch and Hawke was something Flynt might turn to his advantage.

'If Mr Blake here is on the account,' Hornigold interceded, 'then what we have to say will affect him. And if Ned trusts him, that's sufficient for me.' He looked around him and lowered his voice. 'Also, it's my experience that where there is one dissenting voice among a crew, that grew to three, there will be others, and so it would behove us to have a staunch man to watch over us, would it not?'

This was perhaps for Hawke's benefit, for Hesikia Hands had assured Flynt that the remainder of the crew was loyal to the captain, something of which Hornigold might well be aware. Hawke was a landsman, like Flynt, and though he had dealings with men like Thatch, he was not one of them.

As they stepped away from the helm to the port side of the quarterdeck, as far away as they could from other ears, Flynt kept a discreet distance, far enough away to show that he had respect for their privacy but close enough to catch their conversation.

'I'll make it straight and pointed, Ned,' Hornigold began, 'we need you back in Nassau.'

'I'm headed there, but what's the haste that brings you out to meet me?'

'We must convene the island's general council to decide a course of action regarding the Crown's offer of pardon. Word has already been sent to those captains not in port.'

'I've already made my view plain, Ben. I won't accept it.' He held up a hand to prevent Hornigold from speaking. 'I know you have affection for England, Ben; you made that clear when we sailed together, never attacking an English vessel. I have no such scruples, as you know. I wouldn't give a fart in the direction of forgiveness from that fat German king, nor for the other overfed haunches on the thrones of France, Spain or Holland.'

'The tide is turning,' said Hornigold. 'Until now London has ignored Nassau, ignored us, more or less. But the merchants of Virginia and Charleston have made overtures…'

'I can handle the merchants of Charleston,' Thatch said, one hand giving them a dismissive wave.

'But not Richmond and not New York,' Hornigold insisted. 'Or London.'

'Or Edinburgh,' added Hawke.

Thatch gave him a thoughtful gaze. 'These friends of ours turn on us?'

'Not as such, but they see this new policy from King George and his government as something to take notice of.'

'I thought they had control of the government?'

'They are working to get their man back in power, but they are pragmatic. As their new Grand Master told me, times change, and they must recognise that and change accordingly.'

New Grand Master, Flynt mused. Moncrieff, it had to be. As for their man, that might be Robert Walpole, who thought he had Moncrieff as a lapdog but in reality it was the other way around. Walpole had been Chancellor of the Exchequer and First Lord of the Treasury but was now out of favour, though he was doing everything he could to crawl back into the King's good graces and into government. It was likely he had no notion that he was being used by the Fellowship.

'The offer of pardon runs out in a few months' time, Ned,' Hornigold said. 'We must meet in conclave and decide what course to steer.'

Flynt leaned against the rail, doing his best not to show interest, while feigning a close scrutiny of the men going about their duties. Such intelligence would be vital to Woodes Rogers when he finally reached the island, for he would have to know who he might depend upon and who he might not. And yet, despite his best efforts, his gaze kept drifting towards Hawke.

'I won't argue against it,' Thatch said, as if echoing Flynt's thoughts, 'but neither will I argue for it.'

'I will welcome the offer and argue for acceptance.'

Thatch was surprised. 'You will take the King's favour?'

Hornigold sighed. 'I'm tired of it all, Ned. I have made my fortune and it is safely tucked away in the American colonies. It's time to hang up my sword and take my leisure.'

'But to go cap in hand to George's lackeys and plead for forgiveness...'

'Ned...'

'That's not the life we wanted when we left behind the King's navy, Ben. That's not the life we have led these past years...'

'Ned, I've already accepted it.'

Thatch was stunned into silence. Hornigold looked almost shamed over his admission. 'I sailed to Jamaica in January, while you were out at sea, and took the oath. My past sins are wiped clear, I can face the future with unblemished conscience. You can, too.'

'You're Benjamin Hornigold, by God! You were one of the architects of the free port of Nassau. You helped form the Flying Gang. You led us to riches…'

'And I will endeavour to lead you to freedom, Ned. This time, Whitehall and the Lords of the Admiralty are serious in their conviction to rid Nassau of our kind. We either change our ways or it's a noose on the beach for us. I will act as intermediary with Captain Woodes Rogers, but the pardons must be accepted first. We have time, but that time lessens each day. I would beseech you to see sense, Ned. Leave this life behind, begin a new one with your wife.'

Hawke gave a short laugh. 'Which woman, though.'

Thatch glared at him. 'Have a care with your quips, for you and I are not close friends and I will not accept mockery.'

Hawke was unimpressed by the menacing words. 'You don't frighten me, Thatch. You need me to purchase the goods you thieve and dispose of them.'

'Which you do for profit. And there are always others who would do similar.'

'That's true, but it's also true that you have many women who you call wife. Including one in Charleston who you made whore with your men.'

Thatch straightened his shoulders. 'That's a damnable lie, and you know it.'

'A lie is only a lie until it becomes truth by repetition.'

The pistol appeared in Thatch's hands with such speed that even Flynt was impressed. He pushed himself from where he had been lounging against the rail, but Hornigold was already once again acting as peacemaker, laying a hand on Thatch's arm and forcing the muzzle downwards.

'Ned, control that temper, let us not lose sight of the subject at hand.'

'Aye,' Hawke sneered, 'don't let the syphilis that eats at your brain replace your reason.'

'That's another damned lie. I'm not poxed.'

Hornigold again intervened. 'I know there is little love lost between you two, but we are in sensitive times, and must work together to reach some kind of safe harbour.'

Thatch, his face tight with rage, accepted his friend's advice with a curt nod and replaced the pistol in his bandolier. 'Keep him in line, Ben. Why did you bring him anyway? You know there is bad blood between us.'

Hornigold gave Hawke a warning look, then sighed. 'There is another matter we must discuss.'

'And one more pressing than the predicament you and your fellows face,' Hawke sneered.

'That predicament also faces you, Hawke, for you would hang with the rest of us,' said Thatch. 'You might not have gone raiding yourself, but you reaped the benefits of our labours. And I would not depend on those gentlemen of commerce to defend you.'

Hornigold grew tired of the bickering. 'Damn it, Ned, let us not become embroiled in an argument.'

Again, Thatch deferred to the older man. 'What is this other matter?'

'We also came to ensure you were returning to port immediate.'

'I am.'

'Good, for we need you to magistrate.'

'On whose behalf?'

'Mine,' said Hawke. 'And I wish to ensure that the correct verdict is handed down.'

Thatch's expression didn't change. He was well used to such advances. 'And what is the crime?'

'Theft of property and attempted murder.'

'Of whom?'

'Me.'

That brought a reaction, but this time of amusement. 'Damn me, who did you annoy sufficient to merit that? Apart from me, of course.'

'His name, at time of commission of the crimes, was Samuel Bell. His true name I now know to be Gideon Flynt.'

Flynt's breath sharpened and his heart quickened.

Hawke held out a fat leather pouch. Flynt detected the clink of coin as Thatch weighed it in his hand.

'To help you in your deliberations,' Hawke said.

Justice was for sale in Nassau.

23

For a considerable period after they left, Flynt stood on the quarterdeck staring across at Hornigold's ship, which would sail with them into Nassau. Gideon was in custody and his fate had been bought. He had hoped to find him at liberty, help him take back his wife and then flee the island. He had no clear idea how, but then he seldom did have much of a plan in mind, his practice being more to react to and capitalise on circumstance whenever he could. Now he had to first find a way to save Gideon from execution.

But that would have to wait. There was business to be done in the captain's cabin.

As Thatch had said, Hesikia was to lose his leg.

The cook, an Aberdonian by the name of Neil Paterson, had done what he could to save it. He'd tried to halt the bleeding with liniment made from wild pomegranate seed. Hands was placed on a light diet of very little meat. But it was a bad wound. The knee was shattered and, as Paterson observed, no man on God's earth would put it together again. Putrefaction had also set in. To help fight the pain, Paterson had plied his patient with a mix of opium and alcohol. It was called a tincture of laudanum, advocated by Thomas Sydenham, a Dorset physician whose *Observationes Medicae* was highly regarded. Paterson was only a cook but he was an educated one. The quartermaster, however, didn't fully trust what he called a 'heathenish concoction' and put his faith in the soporific qualities of brandy supplied by Captain Thatch. Flynt sat with him as he downed goblet after goblet of the spirit, even sharing a cup but sipping sparingly, for he knew he had a duty to perform once Hands was sufficiently soused.

While they waited, Flynt wished to speak with the quartermaster about the conversation he'd overheard but was unsure of how to broach the subject. He felt he could trust him, but he had been wrong before.

Even though Thatch could be somewhat volatile in his temperament, the goading of Rufus Turner being an example, the quartermaster was loyal to him, so if he queried too intensely concerning what might occur on Nassau during Gideon's trial, there was always the possibility that Hands, when he recovered from the ordeal he faced, might report it. Flynt had sided with Thatch not because he had faith in him but because he had to. Reaching New Providence was even more imperative now, and to fail when he was so close would be a tragedy.

'This be a fine drop,' Hands said, holding his goblet out for Flynt to refill it. That was his fifth draught and he might soon be too incoherent to provide information.

Flynt tipped brandy into the proffered vessel. 'The ship had visitors this morning.'

'Aye, I hears as such,' Hands said. 'Old Ben Hornigold and that rat bastard Toby Hawke.'

That neither Thatch nor Hands held any affection for Hawke was reassuring. 'The man Hawke...'

'Rat bastard,' Hands mumbled over the lip of his goblet. 'Hawke is a barnacle on our hull. He clings there, leeching off us.'

'He is a crimp?' Flynt used the London slang for a receiver of stolen goods. He had listened to the crewmen speak and many of them used idioms with which he was familiar.

'Aye. He buys the goods from us and then sells them in Charleston or Virginia or New York, even London, for a greater price. He'll even christen silver and gold goods so as nobody might suspect they was lifted from the holds of a Frenchie or Spaniard.' He took another gulp of brandy. 'Or English, come to that.'

To christen stolen goods was to remove any identifying marks, like engravings, and replace them with new ones, creating a more palatable provenance for the buyer.

Flynt made sure his voice remained conversational. 'Hawke wishes the captain to officiate at some kind of trial in Nassau.'

Another mouthful of brandy was downed, another grimace of pain from Hands. 'Aye.' He paused, took a series of deep breaths. 'Where be Paterson? This leg be giving me considerable discomfort. Time it were off, I say.'

Flynt wished to press his own agenda through but realised he had to remain casual. 'You have faith in the cook's skill?'

Hands laughed through his agony. 'Ain't got no choice, has I? It's him with his saws and blades, or the poison in this here pin will take me for sure.'

'Has he performed such a function before?'

'Not with no human person. He's cook, been cook for many years, and so has butchered livestock for the pot. He knows his way through flesh and bone.'

That didn't reassure Flynt overmuch and even through his alcohol-hazed eyes Hands noticed his expression. 'Don't you be fretting now, Mr Blake. Paterson, he was apprentice to a navy surgeon back in his youth. He's watched it done and done it himself since. He'll see me right.'

He held out his goblet again. Flynt refilled it, noticed the quartermaster's eyes were beginning to sail. He had little time. He had to push forward. 'Does the captain officiate at many trials?'

Hands nodded, his head loose, his speech faltering. 'Magistrate... he is... for Nassau. Goes back to...' he took another mouthful, '...goes back to... days when old Ben brought us... to... the island...'

Flynt leaned forward, unable now to keep the urgency from his voice. 'Hawke gave him coin to ensure the correct verdict.'

Hands could barely hold the goblet as he tried to raise it. Flynt took it from him, set it on the floor. 'Will the captain bring a guilty verdict no matter what?'

Hands' head drooped to his chest and Flynt had to prop him up against the headboard. 'Hesikia, please, I have to know. Will Captain Thatch announce sentence no matter the evidence or circumstance?'

But Hesikia Hands had passed out. Flynt swore softly but still ensured that the man was comfortable. He would have to discover the procedure from another source, but he knew not who that would be. He considered going directly to Thatch, but decided against it. To show too much interest in something that was deemed not to be his business might arouse suspicion. Even probing the matter with Hesikia was risky, but he reasoned that the man was so insensate through the opium and the brandy that he would not have been aware.

The cabin door opened and Paterson entered, a book tucked under one arm and carrying a canvas bag in the other. He loomed over the bunk and studied the unconscious quartermaster, then announced, 'He's out of it now. Time to get it done.' He nodded across the cabin.

'Clear that desktop yonder and then help me carry him to it. I'll need elbow room to do my work.'

Flynt did as he was ordered and together they hauled the limp quartermaster to the desk, Flynt taking care with his injured leg.

'Nae call for worry, Mr Blake,' Paterson said, his voice strained by their burden. 'He should feel nothing.'

'Should?'

Paterson merely shook his head. Nothing was certain.

They laid Mr Hands down, he groaning a little but was still numbed, and Paterson fetched his canvas bag, which he set upon the desk beside the patient, then took from it a series of implements, some of which were familiar to Flynt from his army days, especially when he had witnessed the surgeons lopping off Colonel Charters' arm after he had carried him near dead from the field at Malplaquet.

'These tools were taken from a Dutch merchantman some time back,' Paterson explained as he laid out the surgical instruments. 'They've proved right useful on board this and previous vessels.'

There was a variety of blades, a short, sharp knife for incisions and a long curved one for dismembering, razors, a large saw for cutting through bone, a set of forceps, a cauterising tool, a curved stitching needle and waxen thread. They all looked very clean, as if they had been scrubbed and scraped most energetically. Paterson was no proper surgeon but he knew his makeshift trade. He also produced another bag containing bandages and linen from which rose the aroma of some kind of vinegar.

'Oxycrate,' Paterson explained. 'Water and vinegar mixed, to form what they call an astringent to bind the flesh of the stump. I soaked the linen in it and dried it in the galley. I've got a lotion made from whites of eggs here, too.' He produced a flask from the bag. 'That will help stem further bleeding.'

Dicken Trelawny and another crewman Flynt knew only as Hans appeared in the open doorway and Paterson ordered them to grasp Hands by the shoulders and chest, saying that they must be ready to restrain him should he awaken. 'Which he might,' he added, 'for this will be serious painful.' He looked across the desk at Flynt. 'Will you be staying?'

Flynt nodded, which pleased the cook. He opened the book, turned to the correct page and propped it up where he could see it clearly. It was an old, well-thumbed copy of a book called *The Surgion's Mate*, its binding loose through overuse, its cover worn and tattered.

'Hold his shank then,' he said before taking a deep breath, and began to unwrap the knee. For a big man his movements were very delicate. A stench escaped when the final piece of material was plucked from the wound, around which the flesh was a livid red, then turning black and also showing traces of green. Flynt was no expert but he could tell the knee and leg had been rendered useless by Turner's ball.

Paterson hefted an incision knife. 'I will have to do this with considerable speed, in mere minutes if I can, so be prepared to restrain him, lads.'

The men pressed their weight on the quartermaster's upper torso and Flynt grasped his thigh with both hands, hoping that Paterson's blade didn't slip and deprive him of some fingers. The smell of the rotting flesh, mixed with the vinegar, was almost overpowering, and he held his breath for as long as he could while Paterson bent over the desk, muttering what sounded like a soft prayer, and then commencing to cut an inch or two above the discoloured flesh.

He hadn't long begun before Mr Hands awoke and began to scream.

Part Three

Nassau

24

New Providence was smaller than he had expected. As they sailed round the tip of a smaller island and into a narrow channel from the northwest, he noted that it had very little in the way of hills, merely places where the land rose gently. Some of it was timbered, but even from the vantage point of the *Revenge* quarterdeck he could see there had been some clearance over the years.

For the main and only settlement on the island of New Providence, Nassau town was not much to look upon. It nestled on the northeastern coast, facing the smaller island – a long tongue of land named Hog Island. On landing at the quayside, he had stared across the blue channel, which bristled with the masts of a variety of vessels, and wondered if wild pigs roamed that spit of land still. Probably not, for where man went, he left death and destruction in his wake.

A fort stood sentinel on a point overlooking the wide harbour but, even from a distance, he could tell it was severely distressed. Run-down though it was, that would be where they were keeping Gideon, for he doubted there was anywhere that offered more security. The sandy beach was almost white in the sun, and made a pleasing contrast with the turquoise of the water on one side and the line of shrubs and trees with thick green fronds dangling from branches on the other. The water was clearer the further it was from the dock, where it was spoiled by detritus discarded from ships and slimed by an open sewer that ran from the township. Makeshift shelters constructed from various native woods and what appeared to be ship's boards salvaged from wrecks dotted the sand, and the warm air was heavy with the aroma of cooking fires around which men and women sat or lay, eating and drinking, mending clothes, cleaning weapons, laughing.

The town itself was in places little more than a collection of poorly constructed huts with roofs that would not withstand a heavy rain.

There were more substantial buildings, but even they showed signs of poor repair, with evidence of fire damage and musket shot. Ditches in the middle of the roadways carried the waste to the sea, and from enclosures behind the houses he heard the snort of pigs and the cackle of hens. The stench hanging over the streets like a pall reminded him of London. This could have been a paradise but wherever humankind went, they despoiled.

As he made his way up a gradual incline in search of the inn in which John Sanders said his sweetheart worked, he thought of Gideon and how to first confirm he was being held in the fort and then to free him. He was but one man. If Cain had been there, it might have been different, but he was many leagues distant. He would have to find a way, and quickly, for the exchange of coin for a guilty verdict meant there would be no justice on this island, no matter how grand they painted their fledgling republic. Here coin was king, as much as it was in London and Edinburgh.

He stopped a sailor in the process of staggering back to the beach and asked where he would find the Mermaid tavern. The sailor fixed a glassy eye on him and waved an unsteady hand behind him and muttered about it being up ahead and to the left. Apparently he couldn't miss it.

As it turned out, he really couldn't have missed it. The tavern was a hulking building of two floors, the wooden slats of its walls unevenly placed, with solid mud forced into the gaps to keep the winds out. A flight of somewhat rickety steps led from the rutted road to a platform that ran the width of the building and to a set of double doors that dangled askew, as if the person who had fitted them had been guilty of sampling too much of the establishment's goods. A crude sign swung over the entrance; it might have been a mermaid, it might have been a unicorn. In fact, it might have been a cat for all Flynt could tell.

He climbed the steps, feeling the wood give beneath his boots, so he clutched a wooden rail for support – not that it would furnish him much security as it too wobbled beneath his hand. A woman with a face so weathered and wrinkled it resembled ancient parchment, and using a battered wide-brimmed hat as defence against the sun, sat in a curious chair that had two curved pieces of wood attached to the four legs. Nearby was a burning lamp, the purpose of which escaped Flynt as the day was bright. She clutched the bowl of a pipe in one gnarled hand, the stem between her lips as she surveyed him.

'Stranger here, I see,' she said.

'I am, Mother,' he said, respectfully removing his hat. 'Lately arrived.'

The woman removed the pipe. 'Do I hear the sound of Scotland in your words, stranger?'

'Aye, from Edinburgh originally.'

She wrinkled her nose in mock distaste. 'It's always a pleasure to hear the sound of the old country, even if it is from an easterner.'

Flynt laughed. 'I hear the ghost of the old country in your voice, too.'

'You do, but so faint I thought I'd lost it a long time since. I'm a westlander by birth.'

'I had gathered such. I won't hold it against you.'

She smiled, displaying cracked teeth, discoloured by too much of the pipe. 'What's your name, easterner?'

He decided to stick with his *nom de guerre*. 'Joseph Blake.'

'Madeleine McRobert. I own this fine establishment.'

Flynt grinned at the irony that coated her words as she waved a gnarled hand around her.

'What brings you here, Joseph Blake? Gone a-pirating, eh?'

'Only as a means to an end, Mother.'

She replaced the pipe and ruminated upon his words. 'Means to an end, now there's a fine phrase to cover a multitude of sins.' She stared at him with one eye half shut, as if protecting itself from the smoke drifting from the bowl of the pipe. 'And I sense sin from you, easterner.'

He was unsure how to reply to that, but she waved anything he might say away. 'Don't fash yourself, friend. We're all sinners here, otherwise we wouldn't be on this island.' She jerked her head towards the inn behind her. 'You think that's a place for prayers? Nay, it's for drinking and womanising, and occasional for bloodletting, but not as much as you'd think, given the nature of this town. Fighting among the men is frowned on by those who rule this place.'

'Captain Hornigold, Captain Thatch among them?'

'You know those fine gentlemen?'

Her tone suggested she didn't think overmuch of them.

'I came in on Captain Thatch's ship,' he admitted.

She considered that, her mouth working the pipe to send up smoke. 'A means to an end, you say?'

'A means to an end,' he repeated.

'And what is that end?'

'At this moment, it's to find a young lady named Mary Bowers. I understand she works the tables here.'

The faint flicker of a smile crossed her wrinkled lips. 'She serves here, aye, that's true. And what business would you have with the lass?'

He raised the canvas bag containing the dead sailor's few possessions. 'I have the traps of a lad named John Sanders, who has bequeathed them to her.'

'He's dead then?'

'Aye.'

Her sorrow was genuine. 'Such a shame. He was a decent lad, for a freebooter. He was killed taking a prize?'

'No.'

He hoped his taciturnity on the matter conveyed that he had no desire to talk of it further. She understood and raised her pipe stem over her shoulder. 'You'll find her inside. Just ask any of the girls, they'll point her out.'

He replaced his hat. 'I'm obliged to you, Mother.'

'And when you're done with her, come sit with me a while and we'll share a bottle. I like to welcome fellow Scots to this place.'

It wouldn't hurt to speak to someone who knew the lay of the land on the island. 'I'll do that, and thank you.'

He pushed the off-kilter double doors open and stepped inside. He had expected it to be cooler within but the truth was that the air was more humid, perhaps thanks to the press of unwashed bodies mixed with the stench of ale, spirits and the roast pig being turned over on a fire in the far wall. A seemingly treacherous staircase led from the edge of the serving counter, which was little more than a handful of rough planks, perhaps salvaged from a beached ship, resting on four aged casks. The tables were placed close together and each was populated by men of all races and colours. The women present were not shy, for there were naked breasts everywhere he looked. These women navigated between the tables as though they were a pirate vessel at sea, eyes ever alert for a prize and prepared to take it. Others were already perched on laps, a few even straddling their customers, skirts hitched, hips grinding in full view of all, though nobody paid any heed. On the floor above was a series of doors from one of which emerged a large man with his head shaved apart from a long tail that grew from the back of his head and was held together with a bright blue ribbon. Behind him, a young woman adjusted her loose-fitting blouse to almost, but not quite, cover her modesty. Flynt assumed that those with coin sufficient could use one of the upper apartments for their business, while those who were less financially endowed could be pleasured at the table. It was so very different from the refined atmosphere of Mother Grady's in London, and he felt a pang of regret that he had left it and Belle St Clair behind.

He stopped one of the girls, who immediately threw back her shoulders and treated him to a smile that would have been welcoming if it weren't so blatantly insincere. 'I would be grateful if you would point out to me a girl by the name of Mary Bowers.'

The smile on her unwashed face became a sneer. 'Believe me, ducks, there isn't nothing she can do that I can't, and better too.'

Her voice was of the English Midlands, Nottingham perhaps, or thereabouts. He had met many people from that locality who had travelled to London in search of work and fortune. Many found the former, though perhaps not employment that would please God-fearing parents, but few found the latter.

'I have some items that belong to her and have been tasked to ensure she receives them,' he said.

Her eyes dropped to the bag in his hand and she reached out for it. 'I can make sure she gets them, lover.'

Flynt jerked the bag away. 'I would rather deliver them myself, thank you, so if you could just point her out...'

The woman sighed and searched the tavern, finally jerking a thumb to the woman Flynt had seen earlier leaving the upstairs apartment, who was at that moment slowly descending the stairway, one hand on the rough-hewn bannister as her eyes roamed around the men below. John Sanders had said she worked the tables and he had naively thought her to be a serving girl. 'There she is, and good luck to you, for she doesn't give as satisfying a tup as I, and that is the truth.'

She walked on, head held high, in search of someone else to pleasure while Flynt weaved through the labyrinth of tables, pirates and doxies to intercept Mary Bowers at the foot of the stairs. She was small and a little plump, her hair tangled and, had it been washed, of a fair hue. Right away she fixed the same smile he'd seen in the first girl and flicked her blouse open a little further.

'You looking for a tupping, handsome?'

She was London-born, he could tell that as soon as she opened her mouth.

'I have something for you, madam,' he said, politely.

She came off the final step and pressed herself closer to him, her hand reaching for the front of his breeches. 'I'm sure you has, my love, but has you the bunce to pay for it?'

He stepped back and held the canvas bag before him like a shield. 'These are from John Sanders.'

The smile faltered. 'Does I know him?'

'He knew you, if you're Mary Bowers.'

'That's me, but...'

'A young sailor who was under the impression you and he were betrothed.'

She laughed and took the bag from him. 'There's many a young sailor who tells himself that him and I are to be wed.' She pulled the drawstring. 'And why does he send you with this to give to me?'

'He's dead.'

Flynt might have said that John Sanders was merely delayed for all the interest she seemed to show. She peered inside the bag. 'What we got in here then?'

She reached in, rifled around the contents and drew out a red silk shirt. 'I knows this,' she said, her voice suddenly softer. 'I remembers this here shirt.'

'It belonged to him.'

She stared at the garment. 'He was a sweet boy, I recalls that now. He was most proud of this here shirt – his Sunday best, he called it, as though he could wear that to a prayer meeting.' She paused, as if she had a fond remembrance of its former owner, then peered into the sack again. 'Ain't much in here to show reflection of a life lived, is it? Some garments, a pair of old shoes...' She produced a small wooden box, flipped it open to reveal a few coins. 'A penny or two.' She reached back into the bag and came up with an old Bible, battered, its cover ripped. She held it in her hand, a little smile playing. 'He couldn't read, but he had this. He said it give him comfort to know the Word was with him. As if God's eye would be on this here stinking place.' She raised her head to Flynt, her voice hardening a little. 'How did he die anyhow?'

'He died because he wished to take possession of a trinket he thought would please you.'

She shrugged, her eyes averted, but Flynt saw them glisten.

'He died with you in his thoughts, if that matters,' Flynt said.

She blinked. 'Doesn't make no never mind to me.' The words were dismissive but her tone was not. He could hear the emotion behind them. 'He did what he did for his own reasons, and if that didn't play out for him, then it ain't my doing.'

Flynt said nothing. She didn't mean what she said. She was a young woman scraping a living in a place far from home. To survive, she had to be unfeeling. At least, she had to appear that way.

A tear broke through and she wiped it away with the back of her hand, giving the moisture a surprised look. 'I doesn't know why this has affected me so. Yes, he was sweet and he treated me proper but he was just another cull. The men what comes in here have three things on their mind, to soil us, to slap us or to save us, sometimes all three. Why that young sailor thought him and me was bound for the altar, I doesn't know, 'cos I never gave him reason to believe it so. In the end, I tupped him regular and he paid me and that's all the story there is. All there is…'

She trailed off, looking at the shirt again.

Flynt spoke gently. 'Were you his first?'

More tears fell and she wiped them away with her fingers. 'I believe I was. I believe I was. These young 'uns, sometimes they convince themselves that it's more than that. I know you must think me a hard bitch…'

'I don't think that at all, Mary.'

She was only a little older than John Sanders, he estimated, so calling John Sanders 'young 'un' might have made him smile had he not known that the life she led had aged her. He knew nothing of this woman, or why, or how, she had ended up in this run-down tavern on a pirate island.

She sniffed, turning from him to quickly sweep away more tears. When she faced him again her voice had adopted a harder crust once more. 'Life is what it is, and we must abide by it. I'm right sorry that this cove, John whatever his name was…'

She knew his name but he played into her toughened persona. 'Sanders.'

A shrug, as if it didn't matter. 'I'm right sorry he's dead, but only because the coin what he give me in the past is no more. That's all he meant to me. Coin.' Despite her best efforts, her voice wavered again. 'If he felt something different, then there ain't nothing I can do about it, is there?' The words were harsh but behind them he heard the young woman she had once been and might wish to be again. 'I mean, is there?' This time it was more an appeal than a query.

'No,' he said, 'there isn't.'

He left her standing at the foot of the stairs with girls leading culls upwards, young boys and girls conveying bottles and tankards back and forth, an island amid a sea of movement and laughter and drinking and men singing. She held the red shirt in one hand and the canvas bag in the other and mourned not just the loss of a soul she never really knew, but also what might have been in another place, another time, another world.

25

Madeleine McRobert was still seated on the platform beyond the doors, the white clay pipe still gripped between her teeth, but must have left in his absence to fetch another chair, one with four legs only, and also a bottle of brandy and two goblets. He dropped himself into the vacant chair, while she poured two hefty measures of the spirit before handing one to him.

'You found Mary then?' she asked.

He sipped the spirit, let its warmth penetrate. 'Aye.'

Madeleine rocked back and forward on her chair, the timbers beneath it creaking a little, the pipe belching smoke as she watched the activity on the street, where there roamed pedestrians, singly and in groups, some aimlessly, some with purpose. It was a colourful spectacle, made all the more so by the bright sunlight.

'That is a curious arrangement upon which you sit, Mother,' Flynt observed.

Madeleine looked down at the curved slats on which the chair rested. 'A gentleman in Virginia made this, based on cribs built in such manner. I find it most restful to sit here and watch the world go by.'

'You have no duties in the tavern to attend to?'

She gave him a stern look. 'You think me slothful?'

'I meant no disrespect...'

Her face crinkled with good humour. 'I'm having some fun with you, easterner. The tavern sees to itself in many ways, and at this hour of the afternoon I usually take my leisure here, a-watching the world go by. Or at least, the world that is Nassau.'

'How long have you been on the island?'

She removed the pipe, stared at the bowl, then tapped it out on the arm of her chair. 'Came here with my husband, God rest his eternal soul, must be, oh, more'n twenty year since. We'd met and wed on Barbados and he thought he'd try his luck farming here, but he didn't make a go of it. He was a passable seaman, so he went on the privateering account, made a sufficiency eventual to buy this place. He's gone now, God rest him.' She stared at the now empty pipe bowl. 'He was a good man, at heart. Never raised a hand to me, never once.' She jerked her head to the tavern doorway. 'Never dallied with the girls yonder neither, which, being a man, and you all being ruled by that chunk of meat between your legs, was something to praise. I stayed on, took over the running of the old Mermaid, made this my home, ensured them girls in there have a roof and a place of safety. Seen the English proprietors and governors come and go, weathered the attacks by the Spanish, watched old Hornigold make it what it is now.' She looked again at the throng. 'Such as it is.'

'When did you leave Scotland?'

She reached within the folds of her voluminous apparel for a leather pouch filled with tobacco and replenished her pipe, then fished at her side for a taper, and Flynt understood the need for the lamp on the floor. She ignited the edge against the flame and set it to the makings, puffing mightily to bring the pipe to life. Once satisfied, she waved the taper expertly to extinguish it and then settled back again.

'It was 1668 when my mother and me left for Barbados. I was but a bairn.'

'Just you and she alone?'

'My father was of the Covenant and had fought at Rullion Green.'

Flynt understood. While in her care, his aunt had taught him much about Scotland's history and he knew something of the 1666 rebellion by the Presbyterians of the west, known as Covenanters, against attempts by the Stuart king to impose episcopalianism on the Scottish Kirk. There had only been one major engagement, at a place called Rullion Green in the Pentland Hills near Edinburgh. It didn't go well for the Covenanters.

'Your father was among the dead?'

Madeleine shook her head. 'No, he was one of the captured.'

There had been a rash of executions following the insurrection, Flynt recalled being taught, a few men going to their doom as a warning against any future uprisings, but he suspected that was not the fate of this lady's father.

'Aye, there's many a tale told of the battle and the blood spilled after, tales that are both tragic and heroic,' she continued. 'My father was one of the lucky ones, I suppose, for he didn't face the headsman in Glasgow or Ayr. He was transported along with others to Barbados, to labour in servitude. As soon as we could, my mother and I took ship to be with him. There was nothing left for us in the westlands of Scotland, and my mother suspected that life there would continue to be unwelcoming to those of our faith, and she was correct, for it was but a few years before the King's troops were shooting innocent folk down in the fields just for cleaving to their belief.' She snorted in derision. 'And the Stuart kings were of Scottish blood, too, yet they turned on their ane folk when it suited them. For the Covenanters it was God, Kirk and King, but the Stuarts didn't take to that arrangement, for they wished to put themselves above the Almighty.' She hawked and spat towards the floor. 'That for them and their like, the old country is well shot of them, even if it does mean a German on the throne.' She sucked on the pipe for a moment. 'You're wondering, no doubt, how a lass raised in the sight of God allows the fornication and wantonness that goes on beyond that door?'

'I wasn't.'

'I'll tell you anyway. I lost my interest in that bastard up above when he took my father. He obeyed his laws till the day he died and all he had in return was suffering, my mother too. I won't bow my head to a god who allows his followers to be so treated while their persecutors prosper. I've seen rich men, landowners, slave owners, mete out cruelty after cruelty while proclaiming love for God. I care nought for them paying in the afterlife, I want them to pay here and now. What goes on in there? That's life. That's the way of it, and if it's damnation ahead of me for being part of it then so be it. What say you, easterner?'

'I'm no believer.' He decided to steer her to matters of more moment, but gradually. 'What's life like in Nassau?'

'It's like life anywhere, we breathe, we eat, we shit, we die. It's a terrible sinful place, thank God, for otherwise I wouldn't make a living. At least here the sun shines and occasional a heavy rain comes and washes it clean, for a while.'

'And the pirate dream of a safe haven?'

She spat out a laugh. 'A republic they call it, but that's a dream, as lasting as the smoke from this here pipe. Ben Hornigold had high hopes for it and for a while it held, but it's breaking down now. Some of these men, they pay lip service to the notion, but in the end they're just greedy bastards who would toss it all away for a handful of pieces of eight, if it suited them to do so.'

'Have you heard of the clemency offered by London?'

'Aye, that's been offered before, didn't come to nothing.'

'I understand this is different.'

'Maybes aye, maybes naw, time will tell. But there's been much talk of it since word reached the island. I tell you, it'll be the death of Hornigold's dream, so it will.'

The conversation between Hornigold and Thatch on board the *Revenge* had already told him there was disagreement. That brought him round to the most delicate of subjects, one he broached with care, for though, like Hesikia, Madeleine McRobert seemed to be trustworthy, he could not be certain.

'There was a man who came on board before we docked,' he began. 'By name of Hawke?'

'Toby Hawke,' she sneered, then dredged up more phlegm and sent it hurtling to the floor. 'To call him bastard would be to do other bastards a disservice.'

'You don't like him?'

'There's few in Nassau that do, but he's what you might call a necessary evil. The raiders can relieve the merchantmen of their cargo but there are many who canna show face in any legitimate port, for to do so would lead them direct to the noose. So Hawke sells it for them, at a price. Of course there are some who can go elsewhere, to the likes of Charleston over there in the Carolinas, maybes, for the governor is not averse to taking a cut of the booty, but they are better at the taking than the selling, so even they will use Hawke's services. So, there's like and there's need and it's the need of him that prevents him from getting his throat slit.'

Flynt took a moment. 'I understand he was nearly killed some years ago.'

'Aye, it weren't long after my man and me arrived here. It was in this very tavern. A sailor took a blade to him, but a devil's poor job he made of it, for Hawke lives still.'

That Gideon had attempted to kill Hawke in this establishment was a surprise. 'This is where it happened?'

'It was Hawke's back then. The seaman crept into his quarters, stabbed him, then made off with some slave or other and her bairn.'

'Any witnesses?'

'None that remain on the island, or lived. It was years ago, and times and towns change. It was maybe two years after that that Hawke sold up to my man and me the building, the land on which it stands, the stock, even what few servants he had. He couldn't abide the place by then. Heard tell that even walking through that door gave him the shits.'

'So he still lives on the island?'

'Between his farm in the interior and his plantation in the Carolinas.'

Another moment. He had to be cautious. 'The sailor who tried to kill him, I've heard he's back on the island.'

'Aye, but Hawke was waiting for him, is what's being said, lured him here, you might say.' She gave him a meaningful look. 'He's from Edinburgh, too.'

'You know this for fact?'

A nod. 'By name of Sam Bell, he said. Sailed here from Jamaica in a longboat, and took himself a room above. At least till he was taken by Hawke's men.'

'Is that so?'

'Aye. My question is, why does this interest you so?'

'I was intrigued by the man Hawke when I saw him. I took an instant dislike to him and wondered why he was entertained by men such as Captain Thatch.'

'A single word covers it. Profit. Each man might not like the other but they make coin from what they do.'

Flynt finished what remained of his brandy, but refused the offer of another. 'So what will happen to the sailor?'

She detected something in the questioning that he had hoped to disguise, for a glint crept into her eyes that wasn't quite suspicion, but certainly curiosity. 'They'll execute him, sure as night follows day.'

That much he had guessed. He knew the answer to the next question but asked it anyway. 'Without trial?'

'Oh, there'll be a trial, but after that they'll march him down to the beach and do him.'

'And I understand the verdict can be bought.'

'On this island, anything can be bought.'

'Will nobody speak for the accused?'

'He will speak for himself, for all the good it will do him. They call this a republic and they talk of democracy and other fine-sounding words that mean nothing to ordinary folk, but in their court it's the same as home. The accuser will have a say but the accused can have nobody at their side.' She still watched him carefully. 'And there's no exceptions, in case you were considering stepping up for the task.'

He managed a half-smile. 'I'm no lawyer.'

'Neither are they, but they'll dress it all up as if they are. And the judge is no judge, but Thatch himself. But something tells me you already knew that, easterner.'

'I'd heard something of that sort.'

'Aye, I'll wager you did.' Her eyes narrowed as she seemed to make her mind up. 'I'd hazard I know what your next question is.'

'And what would that be, Mother?'

'You'll be wondering where they're keeping that poor doomed sailor, I'll be bound.'

He kept the amazement from his face, because that had actually been on his mind, and strove to maintain a casual demeanour. 'It's of no import to me, of that I can assure you. My interest is merely passing.'

'Aye, that's fair enough then,' she said, settling back in acceptance, the pipe clamped between her teeth again, the chair tipping to and fro, the boards protesting against the movement. 'But I'll tell you this – only in passing, of course – they have him penned up in the old fort down by the harbour.'

Flynt had been correct in that then. It was the obvious place.

'And I'll add this,' she said, still rocking, her head resting on the chair's high back. 'Broken-down that fortress may be, but there's a cell below where he'll be chained and no man can free him, not without a key or powder to blow the wall to hell and gone. That sailor will remain there until they drag him to the court.'

'And when will that be?'

'Now that Thatch is back, it will be at his pleasure, but given they like to get their legalising done quickly, for on land they do like taking their leisures, I would hazard that sailor won't have many days left ahead of him...'

26

He took his leave of Madeleine McRobert soon after. He had wished to learn more about Hawke and the island's politics – if the man had any allies at all, for instance – but had tempered his inquiry. She had been welcoming enough, open enough, but he didn't know this woman. Trust didn't come easily to Flynt at the best of times; he had remained breathing thus long only by dint of his suspicious nature, not to mention a hefty dose of luck, and given he was thousands of miles from London, where he was at least comfortable, he had to be doubly careful. Like Moses, he was a stranger in a strange land, and he had to tread carefully, though with some measure of haste, for if Madeleine was correct his father could face execution sooner rather than later. Even if it was somehow delayed, there was Toby Hawke's son, Dan, to consider. Blueskin would have held him for a limited period only. Flynt's passage had been predominantly calm, apart from the storm they met on leaving the English Channel. If Hawke the younger had similar good fortune, then he could only be days from landfall on New Providence.

And then there was the location of Mother Mercy. Hawke had a farm in the island interior, but he knew not where. He also had a plantation in the British Americas, so could have transported her to either of them. He had but a vague idea of the Carolinas' location yet absolutely none as to how distant it was from New Providence, but he could at least reconnoitre the property on the island. He had considered asking Madeleine for its location but decided against it. Her sharp insight as to his true purpose in inquiring into the fort and the trial had unnerved him. He would have to find another way to locate it.

The question was, how? In London he had a network of individuals who provided him with information, young Jack Sheppard being chief among them. Here he had nobody. Hesikia had taken a liking to him, thanks to his support of Captain Thatch, but that would only go so far. The quartermaster held little regard for Hawke, certainly, but it would take Flynt more time to fully gain his trust and that was a luxury he did not have. In any case, Hesikia was in no state to be questioned further, no matter how delicately Flynt might couch it.

With no clear notion as to what to do next, Flynt sauntered back to the waterfront and made a study of the fort. It was run-down, to be sure, but it commanded a good view over the harbour and the channel between New Providence and Hog Island. With decent ordnance, a committed band of defenders and a commander who knew his business, that fort could prove daunting to any invader.

The gate lay open, the doorway itself hanging on one hinge and pushed back, so he edged closer, emboldened by the observation that men and women came and went unhindered. He passed through the threshold into a square compound filled with rubble and a makeshift marketplace, where tables had been set out with foodstuffs and clothing, even a few weapons. The battlements were deserted and Flynt could see only one gun, an ageing cannon lying at an angle thanks to one wheel on its carriage being broken.

A few doorways led off the compound but he had no way of knowing which one might take him to the dungeon Madeleine had mentioned. Nevertheless he began to feel a rise in his hopes, for if the security around that cell was as lax as the fort in general then he might just be able to spirit Gideon away. How he didn't know, but there was at least a glimmer of optimism.

He wandered around the courtyard for at least an hour, searching for a hint that might suggest where they kept Gideon. He contemplated exploring some of the doorways to see where they led, but he held back, for he didn't wish to be challenged as to why he was investigating. He bought a piece of curved yellow fruit from a young woman sitting beside a basket. It was unfamiliar to him and when he found a stool on which to sit, he stared at it, wondering how to eat it. He became aware of the woman watching him and laughing; then she made a motion with her hand that suggested he tear off the top and peel the skin down. He did so and revealed a pale meat which was soft, sweet and pleasing. It also made him realise how sharp-set he had become, having not eaten anything since leaving the ship earlier that morning. The fruit filled a hole and he weighed up his next move.

Gideon was here, somewhere, that he knew. But he was an unfamiliar face, and any inquiry he might make could be viewed with suspicion. Even if he found his father, he had no idea what his next move would be, though he was confident some action would suggest itself to him. He was used to thinking on his feet.

And then his gambler's luck held. Emerging from one of the doorways were Toby Hawke and Captain Hornigold, deep in conversation. Flynt stepped into the shadow cast by the battlements and watched as they walked to the gate. By his expression, Hornigold seemed to be holding onto his patience as Hawke spoke. Once they had vanished, he strode towards the door and found himself faced with two wooden stairways, one leading upwards to the battlements, the other below, the wooden steps vanishing into darkness. He had no idea what he was doing but he knew he had to reconnoitre, so he descended carefully, feeling his way down with his feet, both hands brushing the walls on either side. It was much cooler here, the walls being thick and able to withstand the heat of the Caribbean sun, and he was glad he still wore his coat. In the open he had found it something of a hindrance, but had continued to wear it in order to conceal his weaponry more easily. Not that his pistols would have raised an eyebrow for all the men were armed, but old habits die hard. He reached level ground again, but the floor was peppered with loose stones and sudden indentations into which his feet splashed. Somewhere in the darkness he heard the repetitive drip of water and when he reached out to touch the walls, his hand came away damp. Light bled through the gloom so he made for it and found a small room where a bull of a man sat at a table on which a lamp burned, the sleeves of his shirt ripped to accommodate muscles that bulged as though they were ready to explode. He sliced at a chunk of pork with a knife, and a flagon of some form of liquor sat beside the plate. More importantly, in the corner, seated on the dirt floor and chained to the wall, was Gideon, his head slumped on his chest. Excitement and joy electrified his flesh but he remained outwardly disinterested.

The bull looked up as Flynt appeared. 'And who may you be?'

His accent was French but his English was better than some native speakers.

'Joseph Blake,' Flynt replied loudly, the sound of his voice causing Gideon to raise his head. Shock pulsated through Flynt's body, for his father looked so much older. Worse, he looked frail. His hair was lank and loose, his beard grey, his skin, even in the light of the lamp, pallid. At first his eyes were flat but something kindled within them when he recognised his son. Flynt hoped he still had his wits about him, and that on hearing him use a false name would know to keep silent.

The jailer rose from his chair. He looked to be a powerful individual when seated, but on his feet he was positively overwhelming. 'And what is your business here, Monsieur Blake?'

Despite his dismay at seeing Gideon so weakened, Flynt painted on a smile. He had to present himself as unconcerned. 'I've been sent by Captain Thatch.'

He'd no idea what he was going to do or say, but that lie sprang readily to his lips. The name registered but nevertheless the guard's eyes narrowed as one hand came to rest on a cutlass at his side that looked little more than a dagger against his bulk.

'I have not ever seen you before on this island.'

Flynt moved close to Gideon, his back to the jailer. He raised one finger to warn him to remain silent. 'I recently joined Captain Thatch's crew. On the *Queen Anne's Revenge*.'

'You do not look like a seaman.'

Satisfied that his father understood the need for silence, Flynt faced the jailer once again. 'I'm not. I'm a soldier.'

The man remained unconvinced.

'You can send someone to ask the captain, if you like. Joseph Blake, recruited from the merchant vessel the *Sprite* by Quartermaster Hesikia Hands. I'm sorry, I didn't catch your name, friend…'

'I did not throw it, that is why. Why does old Blackbeard send you to this place?'

Flynt kept his voice light, recognising that the man was laying some sort of trap. 'Ah, I wouldn't let him hear you call him that. You know he doesn't like it.'

The huge hand remained on the hilt of the sword. 'And my old friend Hesikia, how be he?'

Flynt sensed another test. 'He is grievously wounded. I assisted Mr Paterson in the removal of a leg shattered by a mutineer's pistol shot.'

He was gambling that news of the events on board the *Queen Anne's Revenge* had already reached shore.

'My Captain Hornigold said that some landsman came to the aid of Captain Thatch. Would that be you?'

Flynt waved a modest hand. 'I was of but minor assistance, though Captain Thatch was most appreciative.'

The jailer's hand dropped from the sword. 'I am Guillaume Barbeau. You may call me Will.'

'Pleasure to meet you, Will.' Flynt held out his hand but the man maintained his distance. Big he was, but stupid he was not. Flynt grinned. 'You don't trust me, friend?'

'You have not yet stated your business.'

Flynt allowed his hand to drop away but ensured it did not stray near to his pistols. He had no desire to antagonise this Goliath. 'I take it this is the man Samuel Bell?'

'That is what he called himself when he was on the account. It is not his true name, though.'

'What's in a name, eh? It won't matter when he steps into eternity, will it? Captain Thatch is to pass judgement on him soon.'

'After the general council meets, that is so.'

'Quite so. He wishes me to talk with the man first.'

'Why?'

'I'll be frank with you, this man will be found guilty, we both know that. But you must also be aware that there is little love lost between my captain and Toby Hawke. Your Captain Hornigold is not overly enamoured with Hawke either, eh?'

Another gamble, based on fleeting facial expressions, but when Will didn't reply Flynt knew he had won another hand.

'Captain Thatch would have prior knowledge of the crime of which this man is accused,' Flynt continued, not at all sure where he was taking this but committing to it nonetheless.

'Why?'

'He's heard Hawke's version – he would know the other side.'

'Then why does he not come here to himself find out?'

'Come, Will, how would that look? The judge having discourse with the accused before the hearings? My God, man, the proprieties must be observed, even though the outcome is a foregone conclusion.'

The Frenchman seemed to struggle with the word proprieties but he finally understood the meaning. 'Nobody here cares a German's fart about such things.'

Flynt was still thinking on his feet and feared he was not making such a good job of it. 'Very few people care a fart of any nationality about Toby Hawke, am I right?'

Will shrugged. 'He is not well loved, that is the case.'

'Exactly. And Captain Thatch would dearly love to be able to set this man free. After all, he attempted to do what many another man on this island dreams of. And it was all so long ago, so my captain says who really cares now?' He dropped his voice, as if there was anyone else to hear but they three. 'We all know Hawke has bought the verdict, but coin can be returned, eh? And Captain Thatch enjoyed a profitable voyage.'

Will was not prepared to be convinced. 'The charge, it has been brought and my Captain Hornigold says the laws of the republic must be followed.'

'Very laudable, but here's the thing.' Flynt jerked his thumb over his shoulder in Gideon's direction. 'This man may well have a compelling argument to make in his defence and Captain Thatch would appreciate knowing what that would be ahead of the hearing, you understand? That way, he can consider how best to acquit him that would satisfy all in Nassau. He doesn't think much of Hawke but, for the moment, we need him, don't we?'

The Frenchman took all this in for a moment. 'And so he sent you in his stead.'

'He did.'

Will accepted it and waved his hand towards Gideon as he sat back down to resume his meal. 'Then be my guest. Have conversation with him, for it is nothing to me.'

'Ah, see, this is most irregular, I think you'll agree.'

Will sliced off a lump of meat that would choke a horse and bit it through. 'There is no regular, no irregular,' he said as he chewed. 'That man, he is destined for the hemp and there is nothing Captain Thatch can do about it.'

'Well, as I said, my captain seeks a way to avoid that. Should he choose, in his guise as magistrate, to hand down a not guilty verdict, slim though that eventuality may be, then his standing will be of sufficient sturdiness to withstand any furore. Hawke will not be pleased with the outcome, should that be the way the wind blows. Were it suspected that you were in some way party to anything underhand, then Hawke's ire would fall upon you. And, as you know, he is not a man it is wise to anger.'

The jailer stopped chewing as he considered this. 'You say I might be blamed?'

'I'm saying you *will* be blamed. When I inform Captain Thatch of how helpful you have been then I'm sure he will protect you as much as is possible, but you haven't signed articles with him. You're with Captain Hornigold.'

The man nodded.

'And Captain Hornigold is devoted to the laws of this republic, and he may not be so minded to forgive. So perhaps it would be best if you took your leave to satisfy a call of nature. I won't be long with this man. By the time you return I will be gone and nobody will be any the wiser.'

Will thought this over, then rose again. 'Do not touch my food, or my wine.'

Flynt held up both hands in supplication. 'Wouldn't dream of it.'

The jailer stopped before exiting. 'A few minutes is all you have, my friend.'

He stooped to squeeze himself through the doorway and was gone.

'Christ, Jonas, what are you doing here?' Gideon whispered.

'I thought I'd take an ocean voyage for my health,' Flynt quipped, as he hastened to his father's side. 'What the hell do you think I'm doing here?'

They stared at one another for a moment, before Flynt's relief at seeing his father alive caused him to wrap his arms around him. Gideon, shackled though he was, did what he could to return the embrace.

'By God, you shouldn't be here, son, but I'm right glad to see you.'

Flynt stepped back again, feeling tears well. 'And I you, Faither.'

Gideon blinked away his own tears. There would be no further show of emotion. That was their way. But when he spoke, there was a roughness in his voice. 'What of Cassie?'

'She's in Jamaica.' Flynt turned his attention to the chain binding Gideon to the wall and gave it a tug. It was more something to do than any real attempt at dislodging it. 'Young Jonas, too.'

Irritation creased Gideon's forehead. 'I wanted her to come to you for their protection, not to bring her halfway across the world.'

'Did you think I would be able to stop her?' He hauled at the chain again. 'Damn, I think they have fixed this thing to the one part of this fort that isn't crumbling.'

'What's she doing in Jamaica?'

Flynt switched his attention to the padlock holding the chain around Gideon's wrist. It was a barrel shackle, with a screw lock. He might be able to pick it with his dub, but it would take time they didn't have. 'Without doubt seeking a ship to bring them here.'

He stepped over his father and inspected the jailer's table.

'You left them alone?' Gideon's tone was accusatory.

Flynt looked around the small room. 'Gabriel Cain is with her.'

That seemed to mollify him. 'He is a competent man.'

'That he is.' He looked under the table and chair. 'Where the hell is the key for that lock? Does Will have it?'

He was already thinking of how he could disable the big man without killing him, though he would do that if need be. The jailer was only doing his captain's bidding, but he was now between Flynt and his intended outcome and that made him expendable.

'Hornigold keeps it on him,' Gideon said.

Damn the man, Flynt thought, glancing at the doorway. 'We must think of some way to get you out of here.'

The strength that had bolstered Gideon's voice when he admonished his son over bringing Cassie vanished. 'You can't help me.'

'We can try.'

Gideon's head shook. 'I failed, Jonas. I almost had Mercy again. I saw her, but that bastard Hawke was waiting for me.'

'He knew you would come. He sent his son for Cassie and he pursued her to London. Obviously, he was unsuccessful.'

'Did you kill him?'

'No.'

Gideon was disappointed. 'You should have. He's as vicious as his father.'

'Is Mercy on Hawke's farm?'

That surprised Gideon. 'You know of his farm?'

'Of it, not where it is. Is she there?'

'Aye. It's a few miles from Nassau, to the south. He has a handful of slaves, but she won't be there for much longer. He proposes to flit the island for good.'

'Because of Woodes Rogers?'

Another surprise. 'You know of that, too?'

Flynt couldn't tell him of his mission. His work for Charters must always remain hidden. 'It's the talk of Nassau, Faither. I can't help but know of it.'

'Are you really part of this Thatch's crew?'

'It's a long story.'

'I heard Hornigold tell Will that some soldier saved Thatch's life. That was you?'

'As I said, it's a long story. Let's concentrate on getting you free.'

Gideon's head shook. 'You have to save Mercy. Hawke will kill her, Jonas.'

'I'll save you both.'

'Leave me be. But what you can do is get to my traps, I left them at my lodgings…'

'The Mermaid.'

Gideon's eyebrows rose. 'You know this?'

'I met your landlady.'

'She will have my belongings then. Tell her Gideon says all is well.'

'She knows your real name?'

'Aye, we hit it off right away. Us both being Scots can go a long way, and she has no love for Hawke, who tried to take back the tavern after her husband died. But she saw him off.' He flashed an appreciative grin. 'She's a tough old girl.'

'What do you need from your traps, Faither?'

Will's heavy footfalls approached so Gideon told him quickly and on the hearing of it, Flynt had to stifle a laugh.

27

He found Thatch easily. A few inquiries at the beach and on the dock revealed that he was in a nearby tavern, little more than a hut with a roof of large leaves covering tables and chairs resting on the sand. The counter was a heavy sideboard, behind which were crates containing bottles of spirits and two casks of ale. Here there were no doxies offering the momentary pleasures of the flesh. This was a place where men came to drink away any troubles they might have, or to discuss business. There were only two men present. Captain Thatch sat in a corner, the sun cutting between the loosely fitted wall planks sending slashes of light across his bearded face. The man with him sported a nose that curved like that of a bird of prey and a short, sharp beard on his chin, with the area around his cheeks and above his upper lip speckled with stubble. His brown hair was worn long and threaded with grey. His hooded eyes narrowed as Flynt approached and he produced a pistol, which he laid upon the tabletop. Thatch placed a reassuring hand on the man's arm.

'Easy now, Charles, I know this man,' he said. 'This is the soldier I spoke of, the one who put a ball into the craven swine who had a notion to back-shoot me. Joseph Blake, meet Captain Charles Vane.'

Vane sat back, but didn't proffer his hand. 'I thank you for the saving of my friend, Mr Blake,' he said. 'A captain needs good men what has his back.'

'I don't like back-shooters either,' Flynt said. 'And Captain Thatch here was good enough to take me on his account, so that deserved my loyalty.'

Vane's grunt might have been agreement or dismissal, Flynt couldn't tell. 'A God-blasted Scotsman, eh? God preserve me from your race, for you is most truculent and obstreperous. I ain't met one yet what is not prepared to complain about something.'

'We're not all alike,' Flynt said. 'I have met many an Englishman who could moan the face from you at the least provocation.'

Vane pursed his lips, but some humour did creep into those hooded eyes. Flynt had suspected the man was deliberately goading him to see how he responded. By biting back, he had raised his standing.

'What brings you to these waters?' Vane asked. 'You ain't no seaman.'

'I was soldier, as the captain here said. And I come here in search of possibilities.'

A laugh then. 'The most likely possibilities in this life is a watery death or a short drop.'

'I've almost experienced the former, and the latter is present even in London, as judging by the timbre of your voice you would know. However, I will endeavour to avoid both eventualities.'

'*Endeavour to avoid both eventualities.*' Vane seemed to savour the phrase. 'Not only a Scot, but one with some education, by God. There ain't nothing worse than a Scotchman with book learning.'

Thatch intervened. 'Charles, we're all here for our own reasons. I trust this man and that be the end of it.'

Vane's guarded eyes didn't move away from Flynt but the pistol vanished below the table again, his position shifting as he thrust it beneath his frock coat. 'He has the look of a man with something on his mind, so I'll take my leave. You knows my position on the matter we was discussing.'

'I knew that even before we began our conversation. You're an intractable man, Charles Vane, that's why I like you.'

Vane was on his feet, his hat in his hand. 'Intractable. That's a word for educated men, Ned, like you and this Scotchman here. Me, I'm just a poor sailor. But you is correct, I will not budge on this matter, whatever old Ben says.'

'I don't expect you to. I merely wished to hear you say it and hope you will repeat it tomorrow at grand council. I will adjudicate and cannot speak for either side.'

Vane sneered. 'Grand council is a jest, the fancy of a cove what grew tired of our life. Ben may be titular leader on this here patch of sand and rock, but his power, such as it ever was, is loosening. This place is done for, Ned, and it's time we found richer pickings elsewhere.'

Vane said nothing further as he left. Thatch watched him go, shaking his head, waving Flynt into the vacated chair. 'He's a good sailor and an even better raider,' he said, 'but Charles Vane is too single-minded to reason an issue.'

Flynt seated himself. 'You have been talking about the pardons and the arrival of Governor Rogers, Captain?'

'Aye. We have grand council on the morn to discuss the course to take. Ben advocates acceptance and the laying down of our arms. Charles, as you heard, takes the opposite view. They'll both make their argument to the men.'

'And you? I thought you agreed with him.'

'Part of me does. Another part grows weary of this life. Charles is right, change is coming but perhaps we should change with it.' He smiled. 'On the other hand, there are still chickens out there waiting to be plucked, and I would have my fill.' He fell silent as he considered this. 'I'll come down one way or t'other, on that you can be certain, Mr Blake. So was Charles correct – does something lie heavy on your mind?'

Flynt took a deep breath. 'I do need to speak to you, Captain.'

'On what matter?'

'On the pardons and the vote and someone who does not have the best interests of you and your colleagues at heart.'

Thatch laughed. 'The list of those who do not have my best interests at heart is an extraordinary lengthy one, Mr Blake, starting with the colonial proprietors and coming all the way down to any honest seaman we may have robbed or beached.'

'This man is somewhat closer to home. I speak of Toby Hawke.'

The captain's smile dropped. 'What of him?'

Another deep breath. 'First, I must make confession.'

Thatch's friendly demeanour had frosted somewhat. 'I am no priest.'

'Nevertheless, I owe you a truth,' Flynt said. 'About why I am in these waters.'

Thatch stared at Flynt with cold, dark eyes. 'You've lied to me?'

'I have, and I apologise, but I did not know the manner of man you were and my true purpose here was not to be divulged.'

Thatch's voice had become a growl and Flynt felt the menace reaching out to touch him. 'But you will divulge it now?'

'I know you now, better than before at any rate, and I believe you are a man who deserves the truth.'

'Mark me, Blake, I always was. I cannot abide a liar...'

'Again, you have my apologies, but when I explain I hope you will understand my need for such falsity.'

Thatch's breath grew harsh and his eyes hard. 'Speak then, and pray to your God that I don't detect any further prevarication from you, for it will not go well.'

Flynt had formulated the narrative as he searched through Gideon's belongings for the item he now carried under his waistcoat. He had also, with Madeleine's assistance, rifled through a trunk of papers left behind by Hawke when he sold the tavern in search of another document. Now that he was about to tell the tale aloud, he was unsure that it was a convincing argument, but he was committed to this hand and he must play it.

'I am not in the Caribbean in search of new possibilities, nor less am I being hunted by the authorities, as poor Quartermaster Hands believes, though in truth, I am not what you might term an honest man.'

He paused, for some reason expecting a comment or a query, but when none came, he continued.

'I am soldier, that is true, but I'm also thief. I have been party to many a crack lay, I am most dexterous with a dub to gain access to apartments, but my true criminal calling was in the high toby, taking my pickings from the heaths around London. I am also a don with the tats and flats.'

'So you are nothing but a common rogue with uncommon luck at dice and cards. Why then the need to hide it? We are all common rogues here.'

'I didn't hide that, I merely didn't expand upon it. What I hid was my reason for being here. I am on this island for a purpose, Captain. My criminal activities brought me to the attention of a group of individuals who have need of men such as I.'

So far everything he had said was the truth. Now was the point that he must diverge from it.

'What sort of men need a robber of houses and a highwayman?'

'Gentlemen of commerce, Captain, what you might call a Fellowship.'

Understanding dawned on Thatch's dark features. 'For what reason did they send you here?'

'Mr Toby Hawke is the reason.'

'Hawke? He is one of them...'

'Perhaps not for long.'

Moncrieff had ensured that Flynt knew Hawke was part of the Fellowship and had hinted that he was no longer in their favour, or at least his favour. Further, there was the suggestion that he might prefer that the man ceased to exist.

'He has lost their trust,' Flynt explained. 'He had been the conduit between them and the pirate fleet, had he not?'

'Aye.'

'He would deal with their preferred merchants in the offloading of the goods you lifted from captured vessels, correct?'

This was a guess on Flynt's part, and to his relief Thatch nodded his agreement.

'It would appear he has been cheating both you and them, by charging them far higher rates than he passed on to you and your colleagues. He is entitled to a profit, they understand that, but not at their expense. Such enterprise might be applauded by some, and dismissed as mere entrepreneurship, but the Fellowship, much like your brethren, take the view that we do not cheat each other.' He paused before adding the next layer to his fiction. 'He broke their rules once before, years ago, and they sent a man to deal with him, to give him warning that such transgressions would not be tolerated. That man now lies in the dungeon of the fort.'

'The prisoner, the one Hawke said tried to murder him?'

'He was used, as I am, for some of their dirtier business. He was a common sailor, but he didn't try to murder him, it was merely a warning.'

'But he stole Hawke's property, the woman and her child.'

Flynt considered the item hidden close to his chest. 'Did he?'

Thatch's eyes narrowed. 'What aren't you saying?'

Flynt needed to draw him in, like a fish on a line, using Thatch's conspicuous enmity for the merchant as bait. 'I'll come to that, but I will say that Hawke overstepped the mark by taking the woman and luring Sam Bell, or Gideon Flynt, back to the island. He had become a leading innkeeper in Edinburgh, which was useful to the Fellowship for, as you know, such places are fine repositories for information. How many fat prizes have fallen into the hands of men such as you as the result of a casual comment in some tavern or whorehouse in Kingston or Port Royal?'

Thatch shrugged, but Flynt knew he had hit a mark.

'The fact is, the Fellowship has further uses for him.'

'You wish me to acquit this man Bell at the hearing?'

'Aye. Once I have him, I will return him to Edinburgh as quickly as possible.'

'And what of Hawke?'

'He must pay the penalty.'

'That will not go well,' Thatch said. 'He is an arrogant bag of pus and I have little regard for him but we still have need of him, at least for now. And he has paid for the verdict.'

Flynt had to take his story a step further. 'He betrays you, Captain, all of you, and not merely in the matter of the goods he sells for you. While he was in Britain he met with Woodes Rogers.'

Thatch's eyes tightened. 'For what reason?'

Flynt raised both hands. 'He has been informing the new governor of those he can rely on here in Nassau and those he can't. All in order to save his own neck from a noose – and put yours in one.'

'Even if I accept the King's pardon?'

'Put not your faith in princes, Captain. Their sympathies are inconstant, their words and promises just as fluid. That goes for their ministers, and their governors. The Fellowship, however, take pride in their loyalty to those who assist them and even though Hawke has bought the verdict, they will be even more generous than he.'

Flynt felt no shame in lying to Thatch. He had a job to do, to save his father, to find Mercy, and they trumped any other consideration. He cared nothing for this man, who would doubtless have him killed on a whim if it improved or protected his own standing, or if it somehow profited him. At this stage, he wished to sow the seeds of suspicion on ground that was already fertile with dislike of Hawke.

Thatch stared at a point just beyond Flynt as he played in his mind what he had been told. Flynt remained as relaxed as he could, though his hands did not stray far from Tact and Diplomacy. He was unsure whether his ruse had been accepted. If it hadn't, he might have to kill Thatch. It was not an eventuality he would wish for because he was certain he would not get far along the beach before he was brought down, either dead or destined to join his father chained to that wall under the eyes of Will Barbeau.

Finally, Thatch's gaze came to rest on him. 'I will have to think upon this. What will you do in the meantime?'

'There is one service you might do me, Captain. I would require your permission, probably in writing, to make inquiry on your behalf into the crime of which Bell is accused.'

'That would be most unusual. I normally satisfy myself with hearing one side speak and then the other, and any witnesses that I feel moved to allow.'

'This is a most unusual circumstance. I would presume that the majority of the crimes over which you preside have been committed more recently, whereas this was near twenty years ago. I would think it not too unreasonable for you to wish to interview any witnesses yourself.'

'And what witnesses do you think there would be after this time?'

'The woman, Mercy.'

Surprise splashed across Thatch's face. 'The slave? By God, that would be damnably unusual.'

'I agree, not the normal course of events, but it would be most useful if we were to know more of that night and the events leading up to it.' Flynt gave the captain a mischievous smile. 'And besides, think how much it would irritate Toby Hawke.'

Flynt saw that argument hit home, though one final consideration still troubled Thatch.

'This slave...' he began.

Flynt knew what was coming. 'Her name is Mercy.'

Thatch waved that away, for her name mattered nothing to him. 'When I said she was Hawke's property and this Sam Bell or whatever he would name himself did steal her, you made question. Tell me now what you meant.'

Flynt drew a bedraggled document wrapped in oilskin from his waistcoat and slid it across the table towards Thatch. 'I think this will explain all...'

28

Flynt had earlier asked Madeleine McRobert where he could obtain a horse, and she had informed him that she kept a mount in a small stable to the rear of the tavern. He had requested that he borrow it and she had agreed willingly, assuring him that the mare would be saddled and waiting for him when he returned from the beach. She was true to her word, for a tall Palomino was tethered at the foot of the steps. Madeleine herself was still in her chair, rocking back and forwards, her pipe still lit.

'Your parlay with Thatch went well?'

'Time will tell.'

He'd told her what he was about. Gideon trusted this woman and he trusted Gideon's judgement. He examined the horse, running a hand over its flank, stroking its muzzle, the image of Horse back in London coming to mind, along with a pang of regret at leaving her. But then he had left so much behind.

'This is a fine animal. Do I see a trace of the Arabian in her?'

'Perhaps, I know little of such things, though the one who sold her to me said that she was sired through a line of horses brought to the islands by the Spanish. Her name is Flora, after my mother.'

Flynt untied the reins from the rail and placed one foot in the stirrup. The mare stirred a little, unsure of this strange person, but he quietened her by saying her name in a soothing tone and gently caressing her neck. He lifted himself easily into the saddle, adjusted his hat over his eyes and then the pistols in his belt to prevent them digging into his thighs.

Madeleine watched this ritual through a cloud of pipe smoke. 'Have a care, my friend,' she said. 'Yon man Hawke is not a man to cross.'

Neither am I, he thought. 'I thank you, Mother, for the use of your fine horse.'

'You look well on her,' the woman said, a note of approval evident. 'You belong on horseback, I'd wager.'

'I have spent a goodly time in the saddle,' he said.

She gave him a closer look, taking in his black hat with the feather, his dark coat and finally the brace of pistols in his belt. 'I am reminded of scripture.' She closed her eyes for a moment, then recited. 'And I looked, and behold a pale horse; and his name that sat upon him was Death. And Hell followed with him.'

Her eyes opened again and Flynt gave the reins a gentle tug. 'I'm not one for scripture either, Mother, but in Hawke's case, it may prove accurate.'

—

Despite the reason for his journey, Flynt found the ride from Nassau to the island's interior most pleasant. The air, away from the stench running in the central gutters of the town, was sweet with a floral scent wafting from the vegetation alongside the trail. The temperature was most pleasant, not stifling thanks to a refreshing breeze that lifted from the azure waters of the sea and wafted around him; at least once he had wit enough to remove his greatcoat and drape it over the saddle before him. He could grow used to such an agreeable climate. He relaxed into the ride, the faint sound of the sea caressing the sand and slapping against rocks, the insistent cry of a brown-and-white bird circling overhead and the rhythm of the hoofbeats combining to have a calming effect.

Flora was also a pleasure to ride and he felt gratified to be back in the saddle. She was a fine beast, not unlike his own mount in both intelligence and temperament, though she would have been of little use on the heaths. The moonlight trade required the cover of shadow and her creamy golden coat would have been far too easily identified. Horse was black as pitch and so would defy description, for there were many black horses on the streets and alleyways of the city and its environs.

The route required no great effort on the horse's part, for as he had noted from the deck of the *Revenge*, the island was relatively flat. Here and there was a rise but nothing that required much exertion. Thatch had provided him with more detailed directions than Gideon's vague statement of distance and direction and they proved impeccable, for he found the farmhouse easily. Flynt dismounted before turning from the track to the narrower one that led to the two-storey wooden house he could see a few hundred yards away. It was a fine-looking property with neat windows and shutters, a solid roof and a raised platform that ran around it like a wooden moat. Using a patch of bushes as a shield, he retrieved his spyglass from his coat pocket to focus it on the house and yard. Flora waited patiently behind him. The rasp of wood on hard ground caused him to switch his attention to a barn lying to the side and just behind the house itself. One of the tall double doors had been pushed open and a man appeared, pushing a Black woman before him.

The calm induced by the pleasant journey evaporated when he realised that the woman was Mother Mercy.

The man was powerfully built and he propelled her forward with a sharp thrust of one hand, causing her to pitch forward, her hands shooting out to break her fall, but he delivered a vicious kick to her backside, sending her sprawling face down onto the dirt. He bent over her, his lips moving, and she pulled herself to her feet to shuffle out of sight behind the corner of the house, the man following. It was only then that Flynt realised her feet were somehow hobbled and her hands bound before her. Cursing softly, he slammed the glass closed and returned to the saddle, gently urging Flora into a trot. He didn't know what the man intended to do but he wasn't about to let it happen.

As he approached, the door of the house opened and Toby Hawke stepped onto the raised walkway. He watched Flynt approach, his closed fists resting on his hips.

He squinted against the sunlight. 'I know you, don't I?'

'We've met.' Flynt's response came out sharper than it should have, but he was concerned over whatever treatment was being dished out to Mother Mercy behind the house.

Hawke remembered, his voice coated with disdain. 'Blackbeard's ship. The soldier who he had adopted as bodyguard.'

'That's right, and it is Captain Thatch who has sent me here today.'

'For what purpose?'

'To make inquiry before tomorrow's hearing.'

'No inquiry is needed. It's cut and dried.'

'The captain thinks differently.'

'Does he, by God?'

'He does.'

'And with whom will you make this inquiry?'

'With the woman named Mercy.'

'That is no woman, that's property.'

Flynt's fingers tightened on Flora's reins. 'I would still have words with her.'

Hawke's head turned slightly to his right to where Flynt had last seen Mercy. 'She's occupied at the moment.'

'Then I will talk with her while she works.'

Hawke's lips twisted into a smirk. 'She's not working.'

'Then in what way is she occupied?'

That was when Flynt heard the strike of a lash against flesh, and a woman's stifled moan. Flynt jerked Flora's head to his left and had her canter in the direction of the sound.

Hawke ran after him, still on the walkway. 'That is none of your business, sir...'

Flynt ignored him and rounded the house into a small courtyard bounded by the barn and an L-shaped structure in front of which stood a handful of ragtag men and women, their clothes in tatters. The building behind them was in considerably poorer repair than the house. Where the house was painted a bright white, the planks that formed these walls remained bare and rotting. Holes in the roof had been patched with canvas and the windows were without glass or shutters. He took a guess that this was their quarters.

Flynt gave it, and the slaves, only a cursory glance for his attention was taken up by Mercy, who was tied to a pillory, her back bent, her arms outstretched, the ragged gown she wore ripped open to reveal her back, which already bore previous marks of the lash, with one fresh welt rising. Even in that brief, initial glimpse he was aware that she had lost a great deal of weight since he last saw her. The man who had forced her from the barn was shaking a long whip out in preparation for another stroke. Flynt's first impulse was to shoot him where he stood but he forced himself to resist the urge, strong though it was. He had to maintain the pretence of a man here on business. Nevertheless, he felt as if he would vomit, but he had to hide his revulsion, had to appear unmoved by what he saw, for to react in the manner instinct and gut demanded would not further his cause.

Two men emerged from another doorway of the house, both armed. Hawke leaned with both hands gripping the rail of his walkway, his rage colouring his cadaverous features. 'This is not your business, sir,' he repeated.

Flynt adopted as nonchalant an air as he could, even though his own fury twisted within his belly. 'Which is the Mercy woman?'

A smile that was also a sneer curled Hawke's lip and he flicked a finger in Mercy's direction. 'That's her. As I said, she is occupied at present and will not be answering any questions.'

Flynt allowed himself a study of his stepmother, swallowing back the bile that burned his throat, his fingers clenching Flora's reins to prevent them from reaching for his pistols. He would require the application of actual tact and diplomacy if he was to prevent further punishment. 'What crime has this woman committed?'

At the sound of his voice, Mercy had twisted her neck and he tugged the reins to guide Flora so she could see him. She opened her mouth as if to say his name but he cut her off, injecting harshness into his voice. 'I asked your owner, not you, woman.'

He accompanied his words with a swift shake of the head which he hoped was imperceptible to Hawke and his men. Thankfully, Mercy understood and remained silent.

'Crime? I would think that is clear as day, sir,' said Hawke, stepping down from the platform. 'This creature absconded from her rightful owner. That's theft. For that she is being punished.'

The tingle in Flynt's fingers to put a ball in Hawke's face increased. 'I cannot interview her if she is incapacitated by a flogging.'

'I don't believe I will be allowing you to interview her at all.'

Flynt leaned forward to reach into the pocket of his coat to produce the note he'd asked Captain Thatch to pen. He held it out towards Hawke, forcing him to descend from the platform. 'Do you recognise that hand?'

Hawke snatched the paper and quickly read the contents, which stated that Flynt was working on Thatch's behalf and Hawke was to provide him with every assistance in allowing him to talk with whomever he pleased in private. The message was couched in the most courteous terms, the words 'I would be obliged' and 'thank you most kindly' being employed, but it was quite clear that the pirate captain, who signed himself Magistrate Thatch with a flourish, would brook no denial. Hawke looked as if he intended to crumple the paper into a ball but Flynt held out his hand for its return, which the man did, though with ill-concealed grace.

'This is outrageous,' he said.

'Do you intend to defy his wishes?' Flynt carefully folded the parchment and thrust it back into his coat pocket. He may have further use of it. 'He is magistrate, after all, and if he would know more about the events, then he is perfectly entitled to do so.'

'I've never heard of such before.'

'You're hearing it now.'

'It's not the damned Bailey, it's a pirate court. The verdict is certain, for I've paid him good coin.'

'Captain Thatch wishes the proprieties be observed, to encourage all in Nassau that there is indeed order to the law. Who knows, it may even impress the new governor when he arrives. However, if you deny me access to the woman today, then Captain Thatch may wish to speak to her himself, perhaps before, but certainly during, the hearing. If – as I've already stated – she is rendered insensate by this punishment, that would be impossible. It may even delay his judgement.'

Hawke spat onto the ground. 'Proprieties be damned. Blackbeard takes his mummery as magistrate too far.'

Flynt shrugged and wheeled Flora around. 'Very well, I shall relay your words to him. I have little doubt he will have words of his own, which he may choose to deliver in person.'

Hawke caught hold of the horse's bridle. 'I didn't say you couldn't speak to the creature, just that I think it foolishness. The man did attempt murder of my person and stole my property. The case is clear, the verdict certain.'

It would be so simple to slam his foot into the man's face, Flynt thought. A mere raising of the leg was all it would take. He saw it all in the mind, heard the crunch of Hawke's nose, then the report of Tact's bark as he put a ball in him. It would have been most pleasurable, but he knew he couldn't do it. There was the brute with the whip and the two men with pistols watching from the house doorway to consider. The slaves would not move to assist the white men, of that he was sure. He also had to free Gideon from that cell. If he committed murder, though he would not have considered killing these men such a thing, then he would be leaving himself open to facing a similar fate as his father. Thatch's backing would only go so far.

No, he was feeling his way along to be sure, but he had to stick to his decided course of action, no matter how hazy its future progress may be in his mind.

He swung his leg, not in Hawke's direction, but over the saddle. 'I will require somewhere to talk with the woman in confidence.'

'You can ask your questions in my presence,' Hawke said.

'I thought it clear that I would speak with her in private.'

'And I would have either myself or one of my men present. I won't have *that*,' he aimed a finger towards Mercy, 'telling lies about what happened. The man Flynt, or Bell as he was then, did attack me within my tavern, took a blade to me. It was only by God's good graces that his thrust was in haste, or through cowardice, and he missed inflicting a mortal wound by a small measure. But he thought me dead and he took that woman and her whelp and fled the island.'

'And did this woman observe all this?'

'She did, no matter what she might say. She was present throughout the craven attack, which was mounted from behind. And I would wish to know if she does speak lies.'

'Nonetheless, I insist that I speak with the witness alone.'

'Insist? You stand on my land, demanding to speak to my property and yet you make insistence? By God, what has happened to this island…?'

Flynt maintained a level tone. 'Mr Hawke, need I remind you of the captain's instructions?'

Hawke's bluster was cut short and he sucked in air through his gritted teeth. 'Very well, you may use the barn yonder. But my man Bartholomew will remain outside the door.'

Flynt forced a smile. 'You fear I might steal her?'

Hawke gestured to the big man with the whip to loosen Mercy's bonds. 'You are Scotch, like that wretch Gideon Flynt, or Samuel Bell, whatever is his real name. Thievery is in your blood.'

Flynt would gladly see what was in Hawke's blood but he had to delay that particular pleasure. He was here to speak to Mercy and ensure she was as safe as she could be, which on recent evidence was not terribly secure. She had already been whipped and he had interrupted a further bout. He suspected Hawke intended to kill her slowly. However, if he could engineer that Thatch set Gideon free, and with him Mercy, in order to curry favour with the Fellowship, and at the same time implicate Hawke in some form of conspiracy with Woodes Rogers, then all would be well. He had already planted that seed, but turning it to his full advantage remained just out of reach.

On the loosening of her bonds, Mercy wilted a little. Flynt's first impulse was to rush to her aid but he knew that would be unwise. He had a part to play here and he must maintain it.

'Help the woman to the barn, if you please, Bartholomew,' he requested, but the man didn't make a move until he received a nod from his employer. He gripped her none too gently by the shoulders and half carried, half pushed her towards the open barn door. She exclaimed in pain and Flynt almost lashed out but, again, he forced restraint, even though it grew exceedingly difficult.

Bartholomew pushed the door closed behind him, the bottom scraping against the hard ground as it had before, and they were alone in the gloom. Flynt held his finger up to encourage Mercy to remain silent and leaned closer to the door as he heard Hawke's voice.

'Don't move from that position until that bastard emerges. I don't trust him.'

If Bartholomew made any response, Flynt didn't hear it. Hawke then hurled some curses towards the other slaves, telling them to get back to their work before he had the hide off them, too. Flynt gestured that he and Mercy move further into the darkness away from the door. It was a carefully kept space, and the four horses that stood in stalls down one side were evidently well cared for. Flynt recalled Cassie telling him that he cared more for his horses than he did for his slaves. His brush with death at Gideon's hands had clearly not changed his priorities.

'What are you doing here?' Mercy demanded, the pleasant tone of the islands he had come to love broken with pain and fear and anger.

'I followed Gideon, who followed you.'

She already knew that. 'Damn him, I had hoped he would know that was what Hawke wanted. Cassie, she is safe?'

'She's safe.' He didn't have the heart to tell her she had also followed Gideon, along with her son. She was worried enough as it was. Flynt was also concerned, for now that he was closer to her, and even in the dim light of the barn, he could see that she was considerably weakened. 'How many times have you been whipped?'

As if he had reminded her, she reached up and hitched the torn fabric over her shoulders, from which it had drooped and all but breached her modesty.

'Today would have been the third, but there have been other blows struck, both foot and fist. It is his intention to whip the life from me as a warning to the others. And now Gideon will die and it will all be for nothing. At least Cassie is safe.'

'I'll get you out of here, Mother. Faither too.'

'God bless you for the saying of it, and if we were in Edinburgh I wouldn't doubt you, but we are not.'

'There is a plan.' He didn't amplify because he still didn't quite know how it would work, or even what his next step would be. All was improvisational and he reflected on how many times tragedy had followed someone out of his depth stating that he had a plan. Anyway, and on this thought he felt shame, he worried that she might say something if punishment was extended. A thought occurred to him. 'Do you have any influence with the other servants?'

Her head shook. 'I know what you think, that I might be able to stir them to some form of rebellion, but they are so few and are in any case broken people – Hawke and his family, and Bartholomew, have seen to that. Hawke houses the more able-bodied on his plantation in the Carolinas, these here are old and infirm and capable of little more than light household work.'

'The farm produces no crop?'

'None to speak of, not for many years, even before I was able to escape. The soil is poor and he has played it out. These poor people would be of no assistance, even if I could rally them. This is a white man's world. If they mounted even the most minor of resistances then they would be trampled underfoot. I won't risk them.'

He considered the sorrowful state of the men and women he had seen. The notion that they might rise up was a slim one. 'I hope nothing of that sort will be needed.'

'And what do you need of me?'

'Keep the faith. By this time on the morrow you and Gideon will be free.'

'How?'

'I'm working on it.'

He almost added the words 'trust me', but Gideon's admonition would have been relayed to his wife, too.

A fist banged on the door and Hawke shouted, 'How much longer? There's work to be done.'

Flynt shot an irritated glance at the source of the interruption. 'One moment more, if you please.' To Mercy, he said in a hushed tone,

'Whatever I say out there, go along with it. I'll do what I can to prevent any further abuse.'

She gripped his hand. 'I matter little, Jonas. Look to yourself and Gideon. He is important.'

'You are important...'

She reached up and placed her palm on his cheek. 'You are one man against many. And the many here are ruthless and dangerous.'

He pressed his own hand to hers. 'I have my moments too, Mother.'

'Save your father. You would not be the man you are today if it were not for him.'

Gideon was not his natural father, that was true, but many of his teachings – and that of Flynt's aunt who helped raise him – had formed him. He had also learned recently that their minds had worked in parallel. As for the thieving and the killing, that was something he learned for himself, though he was now aware that Gideon knew more of those pursuits than had hitherto been apparent.

'On that we agree,' he said. 'But I have also learned much from your teachings.'

Another pounding from the door. 'I grow impatient,' Hawke shouted.

'Leave us be,' Flynt snarled. 'I'm not finished.'

A few muffled curses was the response. Flynt lowered his voice again, urgency now evident. 'Now listen carefully, for I fear time is running out. Here's what will happen...'

He quickly outlined the plan and her part in it, as Hawke continued to pummel the door. He could see that Mercy had questions but there was no time to answer them for the scrape of the wood against the ground announced that Hawke's patience had finally ebbed.

'Enough of this,' Hawke said, his slim frame etched against the sunlight that flooded into the barn, the much larger bulk of Bartholomew looming behind him, the two men in his shadow. 'You have had time sufficient to learn what you have to.'

Flynt rose from where he had been kneeling and brushed wisps of hay from the knees of his breeches. He wanted to thank Mercy but he knew he couldn't, for appearances' sake. He gave her a brisk nod then joined Hawke in the doorway.

'What did the bitch say?' Hawke asked.

'Very much what you have already stated, Mr Hawke,' Flynt lied. 'The man Bell attacked you and then abducted her and her child.'

Hawke seemed satisfied. 'And you will relay that to Blackbeard?'

'I will make report to Captain Thatch, yes.' He looked back at Mercy, who was still seated on the floor, her head down. 'May I offer a word of advice, Mr Hawke?' He didn't wait for permission. 'I wouldn't beat her further if I were you. If the captain wishes to question her himself and she bears signs of having recently been under the lash, he may come to the belief that her words have been beaten into her.'

'I will do with my property as I please and will take no admonishment from a Scotchman.'

'No admonishment, suggestion only. I happen to know that the captain has a dislike for the lash, having been under it himself while in the navy, so he might disapprove of it being administered, even to a slave. You may accept my counsel or leave it, that is your choice, for it makes no difference to me.'

Hoping that his words had hit home and would give Mercy some measure of respite, at least for now, he climbed into the saddle then rode away without looking back. He wanted to do so, wanted to see Mercy again, but he knew he had to maintain his distance. His head was up, his back was straight, his shoulders set.

Inside he was breaking.

29

The men had begun to saunter down the street outside the Mermaid very early. Flynt had taken Gideon's old room, refusing the offer of female company, but slept fitfully. The tavern itself was a constant babble. Voices amplified by the excessive intake of rum, the occasional flash of anger, the sounds of rutting and moaning from the room next door which appeared to continue throughout the night. He was even able to differentiate between the girls who utilised the bed by the audible level and manner of their passions, which were no doubt feigned, for it was most unlikely that every man, or any, who bought their services provided such a degree of rapture. In all honesty, the sounds did make him regret denying himself such pleasure, for a measure of release from the tension of the day would have been of some assistance, but he knew he would feel guilt afterwards, and he had enough of that on his conscience.

Madeleine prepared for him a breakfast of bread and eggs with coffee which she served to a table that had been laid on the platform overlooking the street. It was from there that he watched the pirates make their way towards the harbour.

'General council,' Madeleine said as she rocked back and forth, her pipe fumes wafting towards him. 'The captains have urged all their men to attend.'

'It is mandatory?' Flynt asked, scooping up the last of his eggs.

'Nay, they have their own mind as to attendance but it has been strongly recommended. A matter of great import, they say, vital to the future of this here republic.' Her smile was ironic. 'Republic, they call it. It never ceases to amaze me how men can delude themselves.'

'It isn't a democracy after all?'

'Another fine word and, oh aye, there is a certain amount of it, but only as far as the captains themselves wish it to be. They've had their quartermasters and mates out dropping a word here, a warning there, a threat or two, a bribe, in order to ensure that their crews vote the way in which they wish.'

Flynt smiled. 'Such is the way with democracy. Votes can be bought with promises, or extorted with threats of dangers posed by opponents, or by inventing or exaggerating another danger. In the end, those who seek power and influence, or who wish to retain it, will say anything, do anything, to achieve their aim. If ever the likes of you or I have the station to have our voice listened to, then we must be very careful in whom we place our faith, for ultimately anyone who seeks such offices must be treated with suspicion.' He picked up his hat from where he had dropped it earlier. 'And with that, I will go now and observe pirate democracy in action.'

'And will you make your voice heard?'

'Perhaps.' He adjusted the hat brim over his eyes against the sun rising over the roofs of Nassau township. 'After all, I've signed articles with the great Captain Edward Thatch and I am entitled to speak out should I feel so moved.'

He followed the growing throng through the broken doors of the fort and found himself a vantage point against a supporting pillar. A platform had been constructed out of crates pushed together over which blankets had been thrown. Benjamin Hornigold was already in position, in conference with Thatch and Vane. Flynt would have dearly loved to overhear their conversation, but he was too far away and it was unlikely he would be granted access. Further men crowded onto the small stage and Flynt presumed them to be the captains of the Nassau pirate fleet. He settled himself against the wooden pillar, his arms folded across his chest, and waited for the proceedings to begin. In truth, he was most curious as to how they would be handled.

Around him swirled a myriad of accents, among which he detected not just the various tonalities of England, Scotland, Ireland and Wales, but also French, Dutch, Portuguese, Spanish and African dialects. The republic was a true Babel, a hodgepodge of tongues and nationalities but, unlike the biblical story, in which God struck at humanity's pride and hubris by splintering their shared language into many, that mixture seemed to meld. For the present, at least. Woodes Rogers was no deity but within months he intended to smite this place by bringing them to heel or scattering them to the four winds.

To his right, he became aware of the men parting to make a path for someone thrusting his way through. Hesikia Hands appeared, a pair of roughly hewn supports under each arm, the left leg of his breeches hanging loose at the knee. His face was gaunt, his colour ashen and the effort of walking appeared immense as he neared Flynt, who reached out to assist him. But the man shook him away.

'I must do this alone, landsman,' he said, his voice stretched by the effort. 'I've made it thus far from the dock, I can make the final two feet to that there pillar.'

Flynt stepped back and the quartermaster reached his goal, where he gratefully rested his left shoulder against the wooden column.

'Should you be doing this at all, Hesikia?'

'I can't be a-lying abed, not while there is important parlay to be had. I must learn to get about on this here sole shank or I'll be no man at all. I thank you, however, for your concern.'

It appeared to Flynt that the quartermaster had shed a considerable amount of weight since he took the ball and lost his leg, even since the last time he saw him, which was only two days before.

'You should have a care not to overdo it, though,' he said.

Hesikia caught his breath and gave Flynt a wan smile. 'The cap'n needs good men here. He'll adjudge the proceedings, as befits his position ashore, but he has his views on this notion of pardons and such, and he expects the crew to follow his lead when it comes to the raising of hands.' He jutted a thumb towards his own chest. 'And this is one Hands that will ensure he gets the support he deserves.'

Hesikia's devotion to Thatch reminded him of the real Joseph Blake's towards Jonathan Wild. Such loyalty might have been inspiring but Flynt suspected that both the pirate captain and the Thieftaker General were cut from the same joint of rotting meat. They could each be personable but there was an aura of deceit that hung around them, while Thatch also displayed that decided imbalance of temperament.

While they waited for the council to begin, the captains still engaging in debate of their own, Toby Hawke appeared through the gateway, flanked by the two men Flynt had seen at the farm. They had a watchful aspect that the pirates lacked. The newcomers to the assembly were not pirates, on that Flynt would stake gold. Former soldiers perhaps, like him; former criminals perhaps, like him; but still men of violence.

Like him.

Hawke made study of the throng and finally found Flynt. His eyes hardened but he didn't acknowledge him. Flynt, however, gave him a courteous bow and the man was forced to return it with a stiff jerk of the head before he gestured to his men to clear the way so he could gain closer access to the platform.

Hesikia noted the exchange. 'You've endeared yourself to old Hawke, I see, landsman.'

'We've encountered one another.'

The quartermaster watched the merchant's progress. 'I swear that fellow could swagger even when seated. He's not one to be trifled with, friend.'

'Neither am I,' said Flynt.

'Oh, that I knows, that I knows...'

Finally, Thatch broke away from the other two captains and faced the gathering. Flynt thought it to number in the hundreds easily, for there was barely space between each man as they packed into the fort's compound. The pirate leader waited until the chatter subsided and finally stilled. He had no need to call for order, no need to draw their attention, for such was his presence that even those who were not looking at the platform seemed to sense that he was ready to begin.

His dark eyes surveyed the faces before him. 'You all know who I be,' he began. 'You all know my function when I am landed here in Nassau town. I am magistrate for them that break our laws, I am moderator for them that has dispute with their brother. For them that likes a scriptural precedent, I am often called upon to be Solomon deciding upon which woman should be mother to a baby.' He gave the men a sly smile. 'I hasn't as yet declared that a baby should be cut in half, but you all know my reputation...'

Only a slight pause was required before the men erupted into laughter among shouted statements that they did, indeed, know his reputation.

Thatch held up his hands to quieten the levity. 'Now, lads, we have ourselves a solemn duty at this here general council. You knows the proposition before us regarding the offer of amnesty from King George over yonder in London. The governor of Jamaica sent his son to read the proclamation...'

A voice cried out, 'Aye, and we fired upon him and set him a-running.'

Thatch joined in the laughter, then said, 'Aye, that we did. But events has moved on somewhat. What many of you might not know is that fat George is sending a new governor to these islands of the Bahamas, and some of you may know him – Captain Woodes Rogers.'

There was a murmur of recognition from some of the men and Flynt heard one man mutter, 'Hell's teeth.'

Thatch quietened them again. 'Now, we have two advocates for each opposing side of this here argument. Captain Hornigold will argue for acceptance of the offer, Captain Vane taking t'other opposing position, which will surprise none of you, for he hasn't kept his views to himself.'

Another laugh. Even Vane smiled, though it did little to relieve the severity of his features.

'So, here's what I says,' Thatch continued. 'I has my own thoughts on the matter, but in deference to the office that you have granted me these past years I shall maintain my silence until it comes time to make a vote, which will be accomplished as is our custom with a show of hands. Them of you what has lost both hands can raise a stump.'

More laughter. It all sounded very fair and above board, but Flynt recalled Madeleine's words, then gave Hesikia a sideways glance. Had he as quartermaster, even in his stricken state, canvassed the crew of the *Revenge* and cajoled men to toe the line?

Thatch ceded the front of the platform to Hornigold, then took a position at the rear, his hands behind his back.

'I will make my position on the matter most clear from the outset,' Hornigold said. 'I am for the acceptance of the pardon, and I know I'm not alone. Captain Jennings has already sailed for Jamaica to take the oath.'

There was no surprise at this, it apparently being widely known, but there were murmurings of distaste, with one man shouting, 'Good riddance to the bastard.'

Hornigold raised a hand for silence. 'I know, he wasn't popular, the Lord himself knows he and I seldom saw eye to eye, but he sees the way the wind blows and I would urge you all to do likewise.' He paused. 'I, too, have accepted the King's amnesty.'

Those men who were unaware of this began to murmur, discussion breaking out between groups as to the advantages and disadvantages of the situation.

'What about what we have built here?' shouted one man. 'This republic of ours? Do we just abandon that what you has helped create?'

'When I first came here men of our kind needed such a haven, a home from which we could raid, it being close to the shipping lanes and within easy reach of the other islands in the Caribbean and the coast of the colonies. And so we banded together, the captains, the crews, and became the Flying Gang.'

Mention of the name of the pirate fleet brought cheers, and Hornigold had to wait until they subsided before continuing.

'But that was when the English government paid us little heed, for they had issues of their own to contend with. They did not look in our direction. Yes, they sent the occasional warship to harass us, but that was mere gesture.' He paused and scanned the faces. 'But that government looks in our direction now. The merchants we have targeted, the governors of the colonies here and on the mainland, have all demanded that they act, and act they have. This offer of pardon to any man who lays down his cutlass and vows never to go raiding again is a solemn one. They seek to eradicate our kind without placing any necks in the noose, but mark me, and mark me well on this, those who dismiss the offer and who continue to raid will find there be nooses a-plenty. We created this here republic out of self-preservation. Now we shall have to abandon it for the same reason. Some of you will know Woodes Rogers, will have sailed with him, so you will know he is not like those sent previous. They were functionaries, men who fought with quill and parchment, but Captain Rogers is no government lickspittle. He won't be intimidated and he won't be bought. And he will bring down the torments of hell on any man who wishes to continue sailing under the black.'

The sailor who had expressed concern at the mention of Woodes Rogers previously spoke up. 'I sailed with him. He's a hard man. I reckons Captain Hornigold speaks true and we should consider most careful what is being offered.'

Charles Vane had stepped forward now. 'And do what? Give up our life, this free-and-easy existence where we answer to no man, no governor, no king? Return to taking the knee and kissing the arse of them what claims to be better than us? Agree to being paid a pittance for our labours?'

'Aye,' cried another voice, 'if we was paid at all!'

Another voice: 'And even then it were often late.'

Vane nodded in agreement. 'Do you wish to return to face the lash and the yardarm dangle if we as much as question an order that we think is wrong-headed? I say no...'

Voices agreed; others disagreed.

Vane raised his own over them. 'I say no! We built this here town after the Spanish had ravaged it. We chased the pettifogging government lackeys out of it. We created our own laws, our own court. These so-called honest men would rob us blind if it suited them, but when was the last time any brother took from another what was not rightfully his?'

He waited for a response but none came, even though men looked to their neighbours with quizzical glances. Flynt surmised that there had been instances. Vane knew it, too.

'And those that did, they were tried before Captain Thatch here and punished, fair and square. Not by the lash or the noose, for that's the English way, the government way, the King's way. They died clean, a pistol ball to the head or a blade to the heart. But hearken to me, lads, if we allow the English government to gain as much as a foothold on that beach then they will bring the cat and the gallows back, as sure as my name is Charles Vane.' He allowed that to sink in before he spoke again. 'Captain Hornigold here said that any man what accepts the pardon will be safe from such punishment as long as he forsake the pirating life, and I believe he be sincere in that. The question I ask is this – can we trust *their* word? How many here received promises from naval officer or rich merchant and was left the poorer for it? How many men here bear the stripes on their back because they spoke out of turn? How many had friends or brothers who went to the rope for no reason at all?'

There were nods and murmurs of agreement. Others remained silent, or shook their heads. Flynt could not estimate how many fell on one side or the other.

'Can't we negotiate with them?' one man asked. 'Keep our republic but promise not to hunt English ships?'

'Aye, there be plenty of foreign vessels just begging to be plundered,' shouted another.

Hornigold stepped forward again. 'Such a pledge could be made, but how many would keep it? Would you, Captain Vane?'

Vane shrugged but every man knew he would not.

Hornigold turned to another. 'And you, Captain England? Would you keep to such a promise not to raid the ships of the land that is your namesake?'

The man to whom he spoke was tall and angular and resplendent in full wig and frock coat. He shook his head. 'I would not. I owe that country nothing.'

Each of the other captains present was asked the same question and there was a mixture of responses; some may even have been sincere in their pledge to curtail such attacks, but even Flynt could tell that in general such a curtailment would not hold. Sooner or later a rich merchantman would come along and, even if flying English colours, would be taken.

Hawke then rapped the head of his cane on the makeshift platform and attention turned to him.

'And what of the honest merchants?' he asked. 'What of their opinion upon the matter?'

Thatch laughed. 'If you can show me an honest merchant in this assembly, then we shall ask him.'

That prompted further hoots from the men, causing Hawke's features to darken. 'Captain Thatch, you have no cause to besmirch my good name in that manner. Especially as you are this day to make judgement upon a matter that I have brought before the courts. Such a comment might suggest your impartiality is compromised.'

Flynt guessed that Hawke was attempting to create doubt ahead of Gideon's trial, just in case he had to fall back on such an argument, but it caused Thatch's good humour to evaporate. 'I'll judge it fair and square.'

'I am not so certain of that...'

'Damn your eyes, Hawke,' Thatch roared and plucked a pistol from his bandolier, cocking it with the palm of his hand. Immediately, Hawke's men drew their weapons, which then prompted members of Thatch's crew around them to produce weapons of their own. Flynt rested his fingers on Tact and Diplomacy.

'No weapons drawn in council,' Hornigold bellowed. 'Damn you all, if we cannot debate issues peaceful, then perhaps Whitehall be correct in their assessment that we should be extirpated.'

Until now the man had been one of reason, but now he showed the manner of a ship's captain, and a pirate captain at that. Hornigold may have tired of life in the Flying Gang but he still had steel to show when needed.

'Ned, put that barker away,' he ordered.

Thatch continued to glower at Hawke, who had stepped back behind his men. Nobody displayed any sign of backing down. Flynt eased Tact from his belt, let it hang loose at his side. Hesikia pushed himself away from the pillar to stand awkwardly supported by one crutch under one arm, while drawing his own pistol.

'God damn you, Ned Thatch, put it away,' Hornigold repeated. 'Mr Hawke, you have your men do the same, otherwise there will be bloodshed. General council is where we come together as one, not to slaughter each other.'

Charles Vane drummed his fingers on his pistols and addressed Hawke's men. 'Put them up, lads, unless you want us all to involve ourselves.'

When they looked around them and saw that they were hopelessly outgunned, Hawke told his men to put their pistols away. Thatch, unwillingly, followed suit and all weapons were lowered and thrust under shirts.

Flynt replaced Tact as Hesikia grinned at him. 'Not something you'd find in that there Whitehall Parliament, eh, landsman?'

'Mr Hawke,' Hornigold said, 'did you wish to make a contribution to the debate?'

Hawke, now that the weapons had vanished, stepped out from behind his men. 'Just this. There are those on this island and beyond who have worked with you…'

'Aye, and made a pretty profit from it,' someone close to Flynt shouted.

Hawke sneered in the direction of the voice. 'As have you crewmen, and your captains and quartermasters and bosuns. You all shared in the spoils and I deserved what little I received…'

Thatch had spotted Flynt and even though he displayed no change of expression, he must have recalled Flynt's words from the day before about the man's underhand dealings. At worst Flynt had lied in order to garner the pirate's support for his plan, such as it was; at best, the claims had been conjecture. The real question was, did the pirate captain believe them?

'But the amnesty offered doesn't extend to men of trade like myself,' Hawke continued, 'so I for one will be vacating New Providence for the more welcoming climate of the Carolinas before this Captain Rogers arrives. Others will do the same, for they have no desire to put their necks on a block. So that means that should you wish to continue in the piratical business then you will have to travel further to offload the booty. I am recent returned from Scotland when news of Woodes Rogers' appointment reached Edinburgh, and it was clear that it is his intention to bring settlers to this island, guarded by the military and naval men-of-war. Nassau as it is now, as it has been for these past years, is doomed and my recommendation is that you all pack your traps, take your vessels and leave sooner rather than later. The dream is over, and you must wake up and face the reality.'

He left it at that and the silence that followed his words spoke volumes. The man may have been disliked, even hated, but he spoke with authority.

'Do you hear that?' Vane said. 'He's a-bringing settlers. And along with them will come priests. And along with the priests will come their moralities, and they will expect every man to change his ways.' He looked around him. 'I see many men here who were liberated from slave ships and who agreed to take articles. When the good men and true come, they will send you lads back to the plantations, for there is profit in flesh.'

Flynt detected some hypocrisy here, for he had been informed that the captains of the Flying Gang had been known to take men and women from slavers and sell them themselves, hand-picking the best of them to draw into the crew.

'With the backing of the pen-pushing functionaries and the military they will stamp out the way of life we has built for ourselves.' Vane injected some sadness into his delivery. 'It was a good thing while it lasted but all good things must end, boys, and I for one would rather up anchor and sail away before that end comes.'

Hornigold cleared his throat and looked to Thatch. 'I think it might be time the matter was put to the vote, Ned.'

'Aye,' said Thatch, 'there has been enough jabbering. All those who be in favour of accepting the amnesty, make it known now.'

Thatch's eyes roamed over a forest of raised hands.

'And those against?'

Another mass of limbs rose perpendicular for him to survey, then with the hint of a smile, he addressed Hornigold and Vane. 'I'd say this here council be torn between the two sides, what say you?'

The two captains could see for themselves that the various crews were split down the middle.

'I suggest we set the issue aside for now,' Thatch said. 'Let the lads here chew upon it for a time and revisit in a few days.' He looked directly at Hawke then. 'We has a court hearing to attend to right now.'

30

Some drifted away, others remained, intent on witnessing their own brand of justice being meted out. As the crowd cheered, this being more entertainment than judicial proceeding, Will Barbeau led Gideon on a chain from the dungeon below and guided him to the platform, Hawke watching his every move with a sly smile curling his lip. This was his moment of revenge and it was as legal as was possible in Nassau. Flynt hoped Captain Thatch was about to disappoint him. He studied the captain as he resumed conversation with Hornigold and Vane, barely glancing at the prisoner, who now faced what remained of the gathering, Will still holding the other end of the chain but remaining discreetly behind him. Or as discreetly as the man could manage, for in the daylight he appeared even larger than he had in the dungeon.

Finally, Thatch turned towards the upturned faces again and the murmur of voices stilled. 'This here pirate court is now in session and the accused stands before you. Sam Bell, or Gideon Flynt, is accused of murder and theft of property and if adjudged guilty then he will pay the penalty for it, as is right and proper. Notwithstanding the subject of the general council recent, this is still our republic and there will be adherence to the articles that we all agreed to. I know you all will have heard of the crimes of which this man stands accused, and will know that this be a most unusual proceeding because they happened near twenty years ago, and in the time I have been your magistrate I haven't never seen the likes of it. Now, who accuses this here man?'

Hawke stepped forward. 'I bring suit against him.'

'The court recognises Toby Hawke, merchant of Nassau and points east and west. We don't do no swearing upon a Bible, don't take no oaths, but we do expect accounts to be true and proper. Now, you all saw this witness and I share sharp words earlier. But that was personal and I will conduct my duties as magistrate with due impartiality, based upon the evidence given, on that you have my oath.' Thatch addressed Hawke directly. 'Bearing the aforesaid in mind, why don't you tell this here congress the whys and wherefores of the case.'

'I thank you, Captain Thatch, and I am gratified that our earlier disagreement will not affect your judgement.' Hawke turned to face the other men, his own guards flanking him. 'You all know me, you know who I am, you know I speak true...'

'That's a contempt right there,' shouted one man, eliciting laughter. Thatch smiled but forced it away and glared at the audience.

'Now then, only them that I instructs to speak can give voice in this here court, so kindly all shut your holes.'

Thatch waved Hawke to continue.

Clearing his throat, Hawke said, 'At the time I was owner of the Mermaid tavern as well as tending my mercantile affairs, both here and in the British colonies. The republic hadn't been formed then, but Nassau was a wide-open port even so and was visited by merchantmen and navy, as well as privateer and pirate. This man, Sam Bell, which was the name he then used, was crew of a ship captained by Captain Prince, who brought his vessel in to be careened. It was hauled up down there on the beach while the hull was scraped and cleaned.'

'Prince was a good captain,' a voice offered.

Thatch reared up again. 'Don't be having me tell you again. You may speak if I instruct you, otherwise keep your goddamned yap to yourself.'

A laugh rippled through the assembly. After a nod from Thatch, Hawke went on. 'During his time in port, Bell saw the slave Mercy, who served in the tavern.'

A voice called out, 'Was she one of your whores, Toby?'

'Damn you, do you not hear me?' Thatch yelled. 'I will have this evidence heard without no interruption.' He drew a pistol and waved it towards the crowd. 'The next man who makes comment without my leave will find it's the last time he utters anything.'

The hidden smiles and winks among the men proved to Flynt that they didn't take the threat seriously. Nevertheless, there were no further interruptions.

'No, she wasn't one of my whores,' Hawke said, 'she was but a serving wench and cook. Though I do think she laid herself down under this man Bell most frequent and that alone was a crime, for she is my property, as is the child she bore having had connection with another of my slaves without my permission. This Mercy was a headstrong creature who I often had to beat some humility into.'

Only Flynt winced at that. Nobody else cared. Even the Black pirates displayed no reaction.

'One night he came to me with a proposition. He was shipping out on the early tide and he wished to buy her and the brat from me. He offered what he said was a fair price and to be sure it was, for he had made himself rich while sailing with Captain Prince. But I refused. I suspected then that he had fallen under some spell she had cast but I would be damned if I would allow her to get away from her rightful position. So I refused his offer, most courteously, and thought it to be the end of the matter. But damn me did he not attack me with his blade. It's only by luck, and his haste and terror, that he did not succeed in doing for me, though it was a close thing. He then took my property and made off. By the time I was found next morning he was out to sea.'

His hand had reached up to stroke his chest, whether consciously or unconsciously Flynt couldn't tell.

'It was not so long ago, when I was in Edinburgh on business for the republic, that I saw him in the street. He was older, most certainly, but I knew the blackguard right off. He was living under another name, Gideon Flynt, and I ascertained his whereabouts and bearded him about his crimes. He laughed in my face, declared that no civilised person recognised this republic as a legal entity and crimes committed here were not crimes at all. Unless those crimes are committed by bastards who sailed under the black and who deserved only the hempen scarf.' He paused to let that sink in as many eyes studied Gideon with renewed interest. Gideon kept his own gaze low but his head did shake a little in denial. Hawke saw this and pointed at him. 'He denies this, but those were his exact words, spoken with such contempt for this place and the men who built it. And so, with the aid of some good men, I managed to take back at least part of my property. The child, grown now, evaded us, as did the man Bell, or Flynt. But there he stands, for in his arrogance he came back to this place, to our place, and intended to steal the slave again, but we caught him and could have cut him down there and then like the vermin he is, but we surrendered him to the pirate council, to be judged by this court. Somebody mentioned contempt earlier, a weak jest at my expense, but there stands the real contempt.' Hawke's finger shot back in Gideon's direction. 'Contempt for what has been built here, contempt for our laws, contempt for you, even though he himself made a fortune thieving and raiding. And so I ask this court to hand down judgement of guilty and have due sentence carried out. He attempted murder on my person and he stole my property and for that his life is forfeit, as dictated by the rules of this republic. Thank you.'

He gave them a brief bow and then did the same to Thatch above him on the platform.

'Smartly spoken,' Thatch said, 'and I thank you for it. Now we all knows the particulars of the events in question from one side, I'll hear testimony from t'other. Accused Sam Bell, or Gideon Flynt, this is your opportunity to put forward your tale.' He gave the men watching a sideways glance and added with emphasis, 'Without no interruption, or by God I'll have someone's blood.'

Gideon looked out across the crowd as he gathered his thoughts. 'As this man's entire testimony was a pack of lies...'

'Damn you, sir,' Hawke sneered.

'No interruptions, said I, and I meant it,' Thatch said, rounding on him. 'You've had your say, now stow your comments.'

Hawke glared at him, then Gideon, but nodded his acquiescence. 'I apologise for my behaviour, it was anger at being named liar by a damned thief and would-be murderer.'

'I'll be the judge of who is thief and who is murderer, when I have heard all the evidence.' Thatch turned back to Gideon. 'Get it said, man, for I haven't eat yet and my belly thinks my throat's been cut.'

'Thank you, Captain. I will begin by asking the previous witness a question, if that is agreeable?'

'It's your testimony, use it as you will.'

Gideon addressed Hawke. 'Who else witnessed this attack of which you speak?'

'Who else?'

'Aye, who was present to see me take a blade to you unprovoked?'

'Nobody, we were alone in my room.'

'So you have no one to corroborate that part of your story?'

Hawke frowned. 'Corroborate?'

'It means to support it,' Thatch offered, with a broad smile.

'I know what it means,' Hawke snapped. 'There were many who witnessed the after effects, to which I nearly succumbed, were it not for the grace of God...'

'I think we should leave the Almighty out of this, for I think neither of us would be in His good graces.' There was a scattershot of laughter and Gideon let it play out. 'So there was only you and I there? Nobody else?'

'No other credible person, just the slave, Mercy.'

'I think we shall hear from this slave,' said Thatch. 'Is she here?'

Hawke was unsurpassed but still looked as if he was going to argue the point. He sighed instead. 'She is beyond the gate. I didn't think it fitting that she attend these proceedings unless called.' He jerked his head to one of his men, who strode through the crowd towards the gate. Everyone waited, a murmur of speculation moving around the compound. Flynt felt something change in the air. Whereas before there was a feeling of levity, now there was expectation. This entertainment had suddenly become more interesting.

He was glad to see that Mercy held herself well as she was led towards the platform, but was clearly in some pain. Hawke had given her another gown, the other having been shredded. Gideon tensed noticeably when he saw her, for he sensed her agony.

Thatch ordered that she be brought onto the platform, which didn't please Hawke. He'd had to stand on the ground, while she was being elevated. Thatch gestured for her to stand close to him and she gave Gideon a longing glance as she passed. His bound hands reached out for her but Will tugged the chain to pull him back.

'D'you know of me, woman?' Thatch asked.

She nodded.

'So you know of what I'm capable?'

She nodded again.

'If you tell me falsehoods, if I even think you do so, I'll be most displeased, do you understand?'

'I will tell the truth, sir,' she said, her head down, and it pained Flynt to see her even feign subservience in such a manner.

Thatch gestured towards Hawke. 'You know this man?'

Her eye darted fleetingly towards Hawke, then dropped again. 'He is Toby Hawke.'

Thatch then took her arm and propelled her to face Gideon. She winced at the movement, the wounds on her back clearly troubling her. 'And what of this man?'

'He is my husband, Gideon Flynt.'

'You married him?'

'In the eyes of God, I did, sir.'

'But you are the property of Toby Hawke – you could not marry.'

'I am not his property, sir.'

Thatch seemed puzzled. 'He testified that you are, that this man did attempt murder on his person and then stole you and your child away.'

'No, sir, that's not what happened...'

Hawke exploded. 'By God, this bitch lies! Am I expected to stand here and listen to a slave tell such lies?'

'You will, sir, and I will decide whether or not they be lies.' Thatch then said to Mercy, 'Tell this court of that night's events.'

'It was Toby Hawke who attempted murder. He tried to shoot Gideon and Gideon defended himself.'

More muttering surged around the men listening. Hawke spluttered something profane and ugly but a glare from Thatch prevented him from speaking further. The merchant's face was rigid with anger.

'Why did Mr Hawke wish to harm this man?' Thatch asked.

'Because he wanted me for himself, sir.'

'But you were his property, were you not?'

'No sir. I never was, sir.'

'Then who did own you?'

'Gideon Flynt...'

Hawke couldn't keep his short-lived silence any longer. 'By God, this is outrageous! These are lies, as anyone can tell you...'

'Can they, Mr Hawke?' Thatch looked out to the crowd. 'Is there any man here who can testify that this woman was Mr Hawke's property?'

Looks were exchanged, heads were craned, but nobody came forward.

'Goddamn this court to hell!' Hawke said.

'Have a care, Mr Hawke, for this court is recognised by all here.'

Hawke aimed a wavering finger at Mercy. 'That is my property!'

'Can you prove it? Do you have the bill of sale that proves such?'

'Not on my person...'

'So you have no proofs as to ownership?'

Hawke's mind was clearly working. 'I know where such proofs will be. I can have someone fetch them.'

Thatch smiled, his hand already reaching into the pocket of his coat. 'That won't be necessary, for I have the bill of sale right here.' He produced the document Flynt had given him the day before, the one Gideon had told him was in his room in the Mermaid. 'And it states that the slave Mercy is the property of one Samuel Bell, bought and paid for on the block in Kingston, one year before the events you described here in Nassau.'

Hawke blustered again. 'It's a forgery!'

Thatch made a show of studying it. 'It's a damned good one if it is.'

Hawke was back to finger-pointing again. 'That cur damn near killed me and took that property! She is mine!'

Thatch waved the paper. 'This here says different. And unless you have a witness present who can corroborate ownership, or documentation, then I have no option but to rule accordingly.'

'My son, he can support my claim.'

Thatch made a show of searching the faces of the men beneath him. 'He is here? Bring him forward then.'

Hawke closed his eyes for a second. 'He is not returned yet from England. I would request an adjournment until he arrives back. It will be any day now.'

Thatch considered this. 'I regret I'm not minded to agree to such delay. We favour judgement swift and supported by fact here, regardless of who be plaintiff and who be accused. In any case, I'm eager to be at sea once more.'

'Damn you, Thatch, you would take her word over mine?'

The captain's eyes clouded. 'I favour most men and women over you, Hawke, but that's my personal thought and not that of my position as magistrate of this here island.'

Hawke addressed Ben Hornigold, who had remained at the rear of the platform throughout the proceeding. 'Will you stand by and let this man steal justice?'

Hornigold took his pipe from his mouth. 'When we appointed Ned here to be magistrate we gave him complete autonomy. There have been no complaints thus far. In fact, I believe near every man here will agree that he has been fair and impartial.'

Cries of assent filled the air. Hawke's barely concealed rage was printed on his face and in the clenching of his fists. Thatch stared at him, an amused glint in his eyes.

'So, what say you, lads?' he said, staring at Hawke, while jerking a thumb towards Gideon. 'Do we find this here man guilty or not guilty of attempted murder and theft of property?'

The roar of *not guilty* was loud and clear. Hawke sneered at the men near to him.

'Then to damnation with the lot of you, for I am done. Captain Vane is correct, this so-called republic of yours is doomed, as are you. I give most of you a twelvemonth before you're all dead and rotting in some festering hole, or under the waves.'

'And what of you?' Thatch retorted. 'Are you not tarred with the same brush? In the eyes of civilisation is it not the case that if you lie down with dogs you wake up a-scratching?'

Hawke had already turned away. 'I'll take my chances.'

He led his men through the crowd towards the gate. Flynt began to follow.

'Where you headed, landsman?' Hesikia asked.

'To make sure Hawke doesn't do something someone else might regret.'

Before he left the fort he saw his father and stepmother embrace. When they parted, Gideon found him and nodded his thanks. Flynt acknowledged him with a small wave.

31

By the time he reached Nassau's main street, Hawke and his men were nowhere to be seen. They'd perhaps travelled to the hearing on horseback or in a carriage so he broke into a run, fearful that he may be correct in his prediction of their next move. His swift progress drew curious looks from others in the thoroughfare, even a few protests as he rushed past, but he ignored them, desperate to get back to the Mermaid, but knowing he couldn't beat them to it. Praying he would be on time.

Sweat soaking his back, he reached the steps leading to the doorway, noting the carriage sitting beside them, dismayed at the rocking chair being empty. He pounded up to the door and pushed through, scanning the room for Madeleine or Hawke, but seeing neither. Mary Bowers was perched on the lap of another sailor and he paused beside her.

'I seek Madeleine. Where is she?' he said.

'Here, mate,' the sailor said, 'this doxy is currently occupied, so wait your turn or find yourself another. There be plenty around.'

Flynt ignored him. 'I asked a question, Mary.'

She pacified her customer by running her fingers through his lank hair. 'It's fine, lover, I'll see to you good and proper just presently.' She addressed Flynt. 'She's gone to her apartment.'

'Alone?'

'No.'

That was all Flynt needed to hear. He pushed through the maze of tables, chairs and people to the room at the rear to which Madeleine had led him the day before. One of Hawke's guardians stood outside, a pistol thrust into the left side of his belt, so he slowed down and smiled in a friendly manner.

'I'm here to see the woman McRobert,' he said.

'She's busy,' said the guard.

'I have an appointment.'

'Busy,' the man repeated, with additional emphasis.

Flynt reached for the door handle with his left hand. 'I'll be but a moment.'

The man grabbed his wrist, which was what Flynt wanted, because he didn't notice his right already drawing Tact and, using his body as a shield to hide the exchange, pressing it into his belly. 'What's your name, friend?'

'Ansel.'

'Ansel? Really?'

'Aye, you wish to make sport of it?'

'Not me. Pleased to meet you, Ansel. Now, let's not disturb the patrons. Release my arm and very gently open that door.'

'And if I don't?'

'Then things will grow very unpleasant, very quickly.'

Ansel may have been initially surprised but he seemed most at ease now. 'In front of these here witnesses? They take a dim view of such things in Nassau.'

'You remember me, Ansel? From Hawke's farm yesterday? I'm here on behalf of Captain Thatch, who suspected your employer might do something rash when the tide went against him at the hearing.'

'I was told to allow no one entry, no matter who it were.'

Flynt fished in his pocket with his free hand to produce the letter Thatch had signed and hold it up to the man's face. 'Can you read?' Ansel nodded. 'Do you see the signature?' Ansel nodded again. 'Then let's go inside and we'll see what Mr Hawke says. He knows me and knows under whose authority I work.'

'You want to see Mr Hawke so badly, you go then, leave me at my station.'

There was no way in hell Flynt was going to leave this man outside. He wanted him in that room where he could see him. 'No, I'm very shy and I'd like you to accompany me.' He thrust the document back in his pocket, then reached behind the man to turn the door handle. Giving Ansel a powerful shove to thrust the door open, he pulled Diplomacy and entered. Hawke's other bodyguard was at the other side of the room, beyond a small bed, with Madeleine's arm twisted behind her back. Her face was contorted with pain, Hawke watching, his smile dropping when Ansel flew backwards into the room followed by Flynt presenting both weapons. The man with Madeleine relinquished his hold and immediately reached for one of the pistols in his belt.

Cold rage frosted Flynt's voice as he trained Diplomacy on him. 'By all means, reach for it.'

The man thought better of it and his hand fell away.

'Are you all right?' he asked Madeleine.

She rubbed her shoulder and gave the man a glare that, were it a blow, would have rendered him senseless. 'I've been treated worse.'

'What's the meaning of this?' Hawke demanded. 'I have business with this bawd.'

'Business that required abusing her?'

'She refused to give me what I wanted.'

'And what did you want?'

It was obvious that Hawke considered refusing to reply, but finally opted for openness. Being under a gun can do that. 'She has something I require and has refused to part with it.'

'And what is it you require?' Flynt already knew what it was Hawke sought but asked anyway.

Hawke again mulled not revealing what it was, but again thought better of it. 'A bill of sale, for the slave Mercy. I believe it was among the papers that I left here when I sold the tavern to this jade and her man.'

'I don't have it,' Madeleine said. 'I told him such.'

'You heard the lady,' Flynt said, 'she doesn't have it.'

'She's lying.'

Flynt's smile was thin. 'Madam, are you lying?'

'I'm not,' she said. 'I don't have the damn thing.'

Flynt knew she spoke the truth because he had burned the document in question the night before. Mother Mercy had commented that there was more of Gideon in Flynt than he knew, even though he didn't share his blood. That they had both thought of having documents forged that proved ownership of Mercy and Cassie showed that. Madeleine had also told him that when she and her husband bought the Mermaid they had taken possession of certain papers, and he had reasoned that it was possible that among them was the original bill of sale, stored away and forgotten about. Sure enough, she had found it in the trunk that lay open behind her. That particular document no longer existed.

'There you are, then,' Flynt said. 'She doesn't have it, so let that be an end to the matter.'

'I'm damned if I will!'

Flynt spoke softly, raising Tact a little further. 'You're damned if you don't.'

Even though he had no real intention of discharging his weapons, the fear that crept into Hawke's eyes was most satisfying. This was a man who was brave when he was backed by other men, who would have them do the physical work, incapacitating whoever they were directed to, and he would perhaps come along later and thrust a blade into the body. Flynt had met many such individuals in the past and could tell them by the very same look of fear.

It was that moment that Ansel elected to earn whatever Hawke paid him. He leaped forward, hand scrabbling for his pistol, but Flynt had predicted such a move and whipped Diplomacy round to club him on the cheek. His partner, no longer under Flynt's muzzle, reached for his weapon but Madeleine swung a powerful blow with her left fist straight to his nose. He rocked back and she followed up by hooking her foot around his ankle and jerking his leg from under him. He reached out to the bed for support, missed and went down. He might have considered rising again but Madeleine had other ideas. She slammed the base of her boot squarely on his face, sending his head snapping back as blood sprayed from his nose.

'Nicely done,' Flynt said.

'You don't run such an establishment as this without learning a thing or two about dealing with men,' she said, then stamped hard on the man's groin, causing him to squeal. Even Flynt winced in sympathy.

Hawke became aware that he was not about to gain the upper hand, but managed to maintain some arrogance. He too had no doubt worked out that Flynt was unlikely to fire. 'I know not what game it is you play, sir, but I smell something about you that is most unsavoury. That day at my farm you lied to me about the slave's testimony. I won't forget that.'

'When you lie to a liar, is it still a lie? Or is it merely truth in an alternative form?'

Ansel was beginning to rise, one hand wiping at the gash Flynt's pistol had left on his cheek. Flynt waved Diplomacy towards his stricken friend. 'Help him up, there's a good fellow. Mr Hawke is about to leave.'

Hawke made no move to assist as Ansel obeyed. 'This doesn't end here, you know that.'

Flynt didn't much care. He'd managed to free both Gideon and Mercy and he would have them safe on a ship as soon as he could. Until then he would cultivate the protection of Captain Thatch.

'Next time you see me, it will not go so well for you,' Hawke warned as his men hobbled towards the doorway.

Flynt let him leave without retort. After being bested, men such as he needed to have that final comment to make them feel relevant. Anyway, it wasn't his expression that Flynt noted, but that of his two men. It was not the last he would hear from them.

32

At Thatch's bidding, Hesikia had found lodgings for Gideon and Mercy in a shack near the beach. On seeing the lash marks on her back, Gideon was intent on visiting Hawke's farm and finishing what he had started those many years before, and it took all of Flynt's powers of persuasion, not to mention Mercy's considerable influence, to convince him to let it go. A mixture of foresight and luck had kept him from execution; they would not be so fortunate a second time.

Once out of sight of pirates and populace, Mercy folded Flynt into her arms and held him tight. Gideon stood by with a soft grin. He rested his hand on Flynt's shoulder. 'Thank you, son. I didn't think it would work.'

Flynt laughed to cover his own emotion. 'I confess, I was unsure myself.'

Mercy stepped back. 'How did you draw that man Thatch to our cause?'

'Lies, Mother, piled upon more lies, and such a construct could topple at any time. I have prevailed upon him to find us passage away from this place as soon as possible.'

'For home?'

'First, for Kingston.' He took a deep breath. 'Where Cassie waits.'

As long as Cain's charm worked its magic and he had delayed her sufficiently.

Mercy frowned. 'She accompanied you across the ocean?'

Flynt felt shame, even though he had little choice in the matter. 'Jonas, too.'

He thought she would chastise him but she didn't. Her frown softened and she smiled. 'Then we shall all be together sooner than I thought. A family.'

A family. Flynt had long thought himself to be a man alone. But he wasn't, and he didn't know how to deal with that.

'You have a wrong to right with her, you know that, don't you?' Mercy said gently.

The old familiar guilt lanced through him. 'I know that. I've done what I can...'

'Aye,' Gideon said, 'you've ensured coin was sent to me to pass along, that is true.'

'Coin for her son,' Mercy said. 'Your son.'

He must have displayed some form of surprise, for she smiled, the first time he had seen it since being reunited with her. 'You think we didn't know that Jonas is your son? You think we were blind in the time before Cassie gave birth? Before she married Rab? You think we were blind as he grew?'

He hadn't considered what they'd thought, but he should have known they would be aware.

'After Rab's death, times were most difficult,' Gideon said, 'for she did so love that man. He was kind and decent and he raised the boy as if he was his own...'

'She wouldn't have wanted me there,' Flynt said.

Mercy squeezed Flynt's hand. 'Nevertheless, you must make up ground with the boy, Jonas. With both of them. They are your responsibility now.'

That made him laugh a little. 'Cassie would say she is nobody's responsibility.'

'That is so, and she is correct, for she is a grown woman and I raised her to know her own mind. But you must be there for her when she needs it. The boy needs a father, as with all boys. You had Gideon. Everything you are, it is because of him.'

Gideon cleared his throat and his voice rasped. 'I ken my blood doesn't run in your veins, but you are still my son and I'm proud of you.'

Flynt took a moment to process his thoughts. He was unaccustomed to moments such as this but he had to say something. 'You were... are... a good faither. You always have been. That... creature... whose blood I do share is nothing to me.' He swallowed. 'I have done many questionable things in my life, you know that. Sometimes I wonder if that is his blood coming through. But the decent things I have done, they come from you, and Mercy, and the mother I never knew.'

Mercy touched his cheek, just as she had done in that dismal barn. 'You are a good man. Gideon is a good man. Sometimes good men must do bad things. But between you both, and Cassie and me, we must... we will... ensure that your son never has to make such choices...'

—

On the third day following the general council Thatch summoned him to the drinking shack, where he sat alone at the same table. The landlord, a small man with the broad shoulders of someone who was used to carrying heavy loads, stood behind the sideboard-cum-counter. Thatch gestured towards an empty chair, its back to the door, but Flynt had no intention of placing himself in such a vulnerable position. He'd noted during his earlier visit that there was no other exit, so he moved the chair a little to the left to enable a view of the doorway. Such knowledge was often the difference between life and death. Thatch noted the move but made no comment. He pushed a bottle of brandy across the table, indicating that Flynt should help himself. It was only nine of the clock in the morning and he was not much of an imbiber, but he poured a small measure into a silver goblet, which he admired for its workmanship.

'Took it from a Dutch trader,' Thatch said as he poured himself a liberal measure. 'We got this here brandy from that raid, too. It's right fine, better than a lot of the bilge water they serve here. The cove who runs this place, he keeps it and the other bottles special for me, the goblets, too. He claims to only bring them out when I'm in port.' He drained the goblet with one swallow. 'We're running out, he tells me, so I have no ways of telling if he's been a-serving it to others, but I may have to find myself another Hollander to fleece.'

Flynt sipped his drink. 'So you will not accept the amnesty? You will continue pirating?'

'It's been a good life for me, Mr Blake. I never took to the strictures of His Majesty's navy. I also suspects that my reputation will precede me wherever I go, ill-founded as it may be.'

Not so ill-founded, Flynt thought, recalling his assertions that he'd had to kill in order to remind men of who he was.

Seemingly reading his thoughts, Thatch said, 'I do occasional let my demon dictate my actions, it is true, and if there is a God and I'm called to His account, then I will have some explaining to do, but many of the deeds they say I committed are fabrications entire. They've had me slaughtering whole ship's companies and defiling women across these islands and the mainland. You heard that slur about me inviting my men to lie with my wife.'

'Lies, designed to paint you villain,' Flynt said, playing along.

'Villain is what I am, Mr Blake, there's no denying it and I am not ashamed of it, for the world is full of villains and many of them have the ear of kings and sup with archbishops and popes. I've killed many a man, to be sure, and I've burned, looted, stolen, blasphemed, wenched, been unfaithful to the marriage bed and struck terror into the heart of many a seaman. So those lies have helped me in my chosen profession. But I have my suspicions that those lies will also prevent me from taking any form of amnesty. They will hang some of us, Charles Vane is correct in that assessment. And my neck would be a prize for them to take.'

He poured another full measure, offered another to Flynt, who declined, having barely touched his first one.

'You seem a solid one, Mr Blake,' Thatch said after enjoying a deep draught. 'No seaman, that be most certain, but I would hazard you might learn the ways of a ship right swift. You are not afeared of doing dirty work, I've seen that with my own eyes, and you have a larcenous soul, you've confessed as much your own self. Hesikia is a good man but is on occasion too squeamish to do what needs to be done. And his want of a leg makes him a somewhat lesser man. I need men of your stripe. Men who will do what needs to be done.' The captain paused before his next words, and when they came it was as if he was unwilling to speak them. 'Men I can trust.'

Thatch looked away then, as though ashamed. Flynt understood him in that moment, for he trusted few and often those in whom he did set his faith proved to be a disappointment. It would not have been easy for a man such as Thatch to put into words that he trusted another. Despite what he had witnessed, despite what he knew of the man, Flynt was strangely honoured.

'I thank you for that trust, Captain, but I must see Gideon Flynt and his wife safely back to Scotland.'

'And deal with Hawke, eh?' Thatch said, not expecting a response. 'I would be happy to deal with the man, if that is what your employers wish, but I'd do it for my own reasons. As far as I am concerned the Fellowship can go to hell, for we shall make our own way.'

The image of Mercy's lacerated back flashed across Flynt's vision. Of Madeleine's pained face. Of Hawke's objectionable features as he smiled at both.

'I thank you, but it is something I must do myself,' he said. There was good in him, but there was also darkness. The Paladin had some bad seed.

Thatch stared at him for a moment. 'Very well, but take my advice and allow some time to pass between recent events and the bastard's demise, for appearances' sake.'

Thatch picked up his hat from where he'd left it on the floor, then stood. The bottle had been near full when Flynt sat down; now a goodly proportion was gone in a short space of time. Yet Thatch displayed little of its effects.

'There is a ship departing for Madagascar in a day or so,' he said. 'Captain John Corrigan is master, Black John, they calls him, and he be a friend of mine, but he's not so well known that he can't call in at Kingston on the way. He will take Flynt and his woman that far, and they can find passage back to England. They shall be safe under my care until he weighs anchor. How does that sound?'

'I am most grateful, Captain. Such an arrangement will be most satisfactory.'

Thatch gave him a brisk nod and left the shack. Flynt remained seated for a few moments, sipping his brandy, considering what he knew of this man. That his name had been unduly blackened may or may not be true, but he did seem to generate loyalty from most of his men. That said, the air of danger about him was palpable.

As he ruminated on this, Hawke's man Ansel and his friend entered the shack and surveyed the interior. It was not a large space so they soon made straight for him. Flynt eased Tact from his belt but kept it hidden beneath the table.

'Mr Hawke wishes to see you,' said Ansel.

'Then he may come and see me.'

'He wishes to see you at his home.'

Flynt raised his goblet to his lips with his left hand. 'Please thank him for his kind invitation, but I am quite comfortable here. I would offer you a drink, but it's Captain Thatch's personal stock and he would not thank me for sharing it with any Tom, Dick or Ansel.'

'We ain't here for no drink,' said the other man.

Flynt made a point of studying him. 'I didn't catch your name the other day.'

'Not that it matters, but it's Cornelius.'

Flynt thought he could detect a faint accent. 'Of Low Country stock, are you?'

'My father was from there. I was born here.'

'How's the old tackle today? Still bruised after being paid attention to by a woman, but not in a pleasurable way? Mind, I'd hazard that it's the closest you've come to a female touch in a long time for which you haven't paid, eh?'

Cornelius sneered and his hand twitched to his pistol. Flynt quickly brought Tact into view, his left hand producing Diplomacy. He crossed his forearms so that one was aimed at Ansel, the other at his friend, then sat back in a relaxed manner.

'Have a care, for I am exceeding nervous and the most minute of shocks might make my finger jerk.'

Ansel shook his head towards his companion. 'We ain't here for no trouble.'

'I'm gratified to hear that.'

'We have been sent to fetch you to Mr Hawke, is all.'

'I will not be fetched, thank you. I say again, if Mr Hawke desires my company then he can come to me.'

A memory of a similar conversation with Blueskin Blake a few years before made him smile, for here he was reliving it while using his name. The vagaries of fate merging with the occasional predictabilities of his life and occupation, amused him. The men exchanged glances, unsure what to do. Flynt, tired of the exchange, rose. It was his intention to leave first, for he was not about to allow them the opportunity to lie in wait outside. His sudden movement caused them to step back, fingers automatically closing on pistol butts. Behind them, the owner of the shack had already produced a musket of indeterminate vintage. 'We won't have any trouble here,' he said, his Irish accent as rough as a broken-down cobbled road.

Flynt made a show of replacing his weapons, but he did so very slowly. 'You heard the man,' he said to Ansel.

'It's in your interests to come with us,' Ansel said, his hand falling to his side once more, though Cornelius remained on the alert.

'In what way?'

'Mr Hawke has something to tell you. Something what will interest you.'

Something in the man's smirk caused an alarm bell to clamour in Flynt's head. 'What would that be?'

'You has to come with us to find out, don't you?' Ansel raised his hands. 'You have no need to fear us.'

Cornelius smirked. 'Not you, at least.'

Flynt didn't like the smirk, didn't like the sharp look from Ansel that caused it to wither and die. 'I'm relieved to hear that. However, you may tell Mr Hawke that I will attend him when I have the leisure, and in the meantime, I would be grateful if you two would remain here after I leave for at least five minutes. My nervousness has increased manifold, and who knows what that would do to my trigger fingers?'

'Mr Hawke, he won't like it if we returns empty-handed.'

Flynt smiled. 'I cannot help what Mr Hawke likes or dislikes, but that's the way it will be, for I have business elsewhere.'

He backed away as he spoke, ever alert for either of the men to revert to their true nature and attempt force. The landlord continued to level the musket, but not at him. Perhaps his society with Thatch had endeared him to the man.

'Why not take a seat, boys,' he said, waving the muzzle towards the empty chairs. 'I'll treat you to a rum for free and gratis.'

Ansel seemingly noticed the landlord for the first time, then with a final glance at Flynt, who was almost out of the door, nudged his friend with the back of his hand and indicated that they should be seated. Cornelius did as he was told, but his face revealed that he didn't like being bested once again.

33

Flynt walked as swiftly as he could along the beach towards town. He knew they would come after him and even though Nassau was not as labyrinthian as London, where he might more easily have avoided their attentions, he had made it his business to become better acquainted with its streets and what might be called alleyways, but were merely gaps between the crudely built and more solid structures. He had no intention of leading them to Gideon and Mercy, even though they would, with the minimum of inquiry, trace them.

He was more concerned for Madeleine's wellbeing. That little snigger of Cornelius and his comment that suggested someone other than Flynt had need to fear them had sent an alarm clamouring. He had to ensure she was safe and well, so he made a circuitous route to the inn, staying clear of the main street and keeping to the rear of the houses, where the pigs and chickens roamed, and thus it took him longer than usual to reach the Mermaid.

Madeleine was not in her chair on the platform, so he ensured his pistols were within easy reach, and his sword loosened within the silver cane, before entering. The tavern was doing what he now knew to be its customary trade. A small fortune in pirate coin was exchanged for liquor and wench, and he could understand why Hawke had wished its return. He weaved between the tables to the rear of the tavern, his instincts telling him that something was amiss. He could not explain why he thought so, it was just something that clouded his mind. He knocked on the door to Madeleine's room and waited, but there was no instruction to enter. She could be sleeping and hadn't heard his knock but the woman didn't strike him as someone who required a nap. He knocked on the door once more, but again no sound came from within. He looked around but none of the patrons or the girls, or the serving staff, paid him any heed. Sliding Tact free, but hiding it with his body, he pushed the door open.

She lay on the cot. She might have been sleeping. But she was not. A pool of dark red spread through the blankets beneath her and Flynt could see the gaping wound at her throat. Someone had killed her and then arranged her body, her arms across her chest, her dead eyes staring at the door as if waiting for Flynt to enter. He replaced Tact in his belt as he crossed the room to touch her flesh. Warmth remained, and the blood had not hardened, it was viscous, which meant she had not been dead overlong. Her features had smoothed in death, revealing the ghost of the young woman she had once been. He had brought this upon her. Had he not befriended her, she would live still. The least he could have done was to stay nearby to protect her from the repercussions of the clash with Hawke and his men. He knew they were responsible, perhaps Cornelius only, the one she so easily mastered. His pride would not have taken well to being so shamed.

'I'm so sorry, Madeleine,' he said, gently closing her eyes. More death. It was all he seemed to leave in his wake.

A movement at the door made him whirl, hand clutching for a weapon, but Ansel and his friend already had theirs levelled.

'Ah-ah,' Ansel warned.

Flynt allowed his hand to fall by his side. 'Who did this?'

Cornelius adopted a sly smile, and in that moment Flynt knew for certain it was he.

'Why, you did,' said Ansel, innocently.

So that was the way of it. He was to be blamed.

Ansel moved into the room, but kept his distance. 'Why did you kill her, Blake? Wasn't she your friend?'

Flynt stared directly towards Cornelius then slid his gaze towards Ansel. 'Did you help? Did you hold her down?'

His expression remained innocent. 'I have no idea what you're talking about. We knows you done it. We ourselves sees it, did we not, Cornelius?'

'We did. It is a pity we did not arrive sooner, for we might have saved the poor woman.'

'As it be, dead is dead, and here you is a-standing over her, with the blade of your knife stained with her claret.'

Flynt was momentarily puzzled, but then he saw Cornelius holding up a bloody knife. His fingers itched to draw Tact and wipe the smirk from his face but he would never clear it from his belt before one, or both, killed him. He had to satisfy himself with a malevolent stare. 'I have killed many men in my day,' he said. 'There are few in which I have taken pleasure. I'll enjoy killing you.'

'Oooh,' Ansel said, with a slight laugh. 'That is proper scary, that is.' He hardened his voice. 'Let's have them barkers. Just drop them on the bed there and step aside, if you please.' He sharpened his aim as Flynt's hands moved to his weapons. 'Nice and slow, if you please, like you was stroking a doxy's skin.' Flynt did as he was told, but retained his stick. 'Now step aside a piece, and allow Cornelius here to pick them up.'

Again, he did as ordered and Cornelius eased forward to fetch the weapons.

'Why kill her, Cornelius? She meant you no lasting harm.'

'She was a wizened old bitch,' Cornelius replied. 'She'd lived too long already.'

'There's a lot of that about,' Flynt said, then spoke to Ansel. 'What now? To the fort to face trial?'

'Nah, I told you that Mr Hawke wished to have converse with you and that's what will happen.'

So that was it. They would say he killed Madeleine and was killed going after Hawke.

'Don't be trying no slippery stuff,' Ansel warned, 'for you ain't the only one with a nervous trigger finger.'

Ansel stepped back and waved the pistol towards the door, then he and Cornelius followed at the rear. Nobody paid them any heed as they left the tavern.

34

They kept him on foot, riding their own mounts behind him at a distance should he, as Ansel had said, try anything slippery, but close enough to put a ball in him. They made a curious spectacle as they left the town and, even though the interest of the tavern's patrons had not been piqued, they did draw a number of curious looks in the street, though nobody made inquiry. This may have been a republic, it may have had its laws, but it was still an outlaw haven and as such a man's business was his private affair. Thankfully they hadn't bound his hands, so he at least retained some dignity.

The heat had risen higher than ever before, though it may have merely been because he was walking. He took off his hat and wiped the sweat from his brow with the back of his hand, his other holding his cane. They hadn't taken that from him because they had no idea what lay within its sheath. He would seek an opening to make the most of that oversight.

He was curious as to what Hawke had to say to him. However, he would have preferred it to be on his terms and not by force. As he walked through the countryside, the air sweet, the sun beating on his back, the cry of the mockingbird above, he pondered what action to take. That he could not allow himself to be escorted to the farm was a given, for he would never leave there alive. He shot a glance back at the two men, but they had ensured the gap between them never shortened, so he had little opportunity to surprise them. He could feign some form of malady, perhaps, collapse to the ground and when one came to inspect him, take him down. But then there was the other to consider. Cornelius seemed somewhat dim, but Ansel was no fool. He would make sure he held back, his weapon at the ready.

A grey lizard skittered across the track, its curled tail weaving like a rudder in the dust. Its movement caught Flynt unawares and brought him up short, sparking a derisive laugh from Ansel. 'Scare you, did it? A little creature like that? Perhaps you ain't the bogeyman that you likes to think, eh? Perhaps you ain't so scary after all.'

That brought a snigger from Cornelius and Flynt shot him a sour look. 'You'll discover soon enough how scary I am.'

Something in that look killed Cornelius's laugh, prompting Ansel to snort. 'We shall see, bogeyman, we shall see. Now, let's pick up the pace, eh?' He rubbed his belly. 'I am right sharp-set and yearn for some pork and bread.'

Flynt began walking again. His threat to Cornelius was little more than bluster, he knew that. The annoying thing was, Ansel also knew that. He couldn't conceive of any way to get close enough to them to do any damage without himself catching a pistol ball. His gambler's luck had deserted him, it seemed.

But then...

'Gentlemen, I wonder if you can help me.'

As soon as he heard it, Flynt knew his luck had not vanished, it had merely taken human form. He turned just as Ansel and Cornelius twisted in their saddles to face the source.

Gabriel Cain stood in the centre of the track behind them, a pistol in each hand, and Flynt couldn't keep a wide grin from spreading across his face. Cornelius was not so pleased, for he automatically reached for his own weapon but Cain raised the one in his right hand.

'I wouldn't,' he said with his customary smile. 'I'm frightful warm, I've been walking for a long time, a practice of which I am far from fond, and even though I might attempt some precision shooting to merely incapacitate you, I fear my aim would be a little off and the result would prove lethal. That would cause me considerable distress, for I have nothing against you fellows.'

Ansel had recovered well enough from the surprise to tug his horse's reins to round fully on Cain and lean his forearm upon the pommel of his saddle. 'Then why do you come upon us with pistols cocked and raised?'

'I would relieve you of that man there.'

'Why? What is he to you?'

'He is my friend, and I can't bear to see him so distressed in this heat. And afoot, too. That's no way for a gentleman to travel.'

'He ain't no gentleman, he's a Scotchman.'

'That's a good point, and extremely well made. But nonetheless, I will not allow him to be so treated, so if you will give him up to my custody, I would be most appreciative.'

Ansel seemed amused by the situation. 'And what good would that do, for you would both still be walking.'

'Ah, not if you and your friend here gave up your mounts.'

Ansel laughed. 'Why would we do that?'

Cain smiled. 'Christian charity?'

Cornelius felt bold enough to speak. 'You expect us to hand over our prisoner and go afoot ourselves? Have you been touched by the sun, stranger?'

Cain squinted at the sky. 'Very likely, for it is devilish hot.' When his attention reverted to Ansel, the good humour in his voice was gone. 'But if you have no charity in your hearts, then let me put it this way. You either relinquish my friend from your custody and depart upright or we leave you here prone in the dust. The choice is yours.'

Silence then as the two men played the staring game with Cain. Flynt softly twisted the handle of his cane and slid the blade free while taking one noiseless step closer to the horses, then two, then three. Ansel and Cornelius didn't notice.

'So what is it to be, gentlemen?' Cain asked, his tone once again light. 'Do we trade? Your lives in return for my friend there and those two fine beasts?'

Ansel chose that moment to reach for his pistol, even managed to clear it from his belt before Cain's shot plucked him from the saddle. Cornelius was faster in drawing but his shot failed to find its mark, perhaps because his horse reared at the sound of the gunfire. Cain was about to fire his second pistol but a head shake from Flynt made him lower it. Cornelius, still trying to bring his horse under control, fumbled for his second pistol but Flynt closed fast and hauled him from the saddle to pitch him onto the ground. Cornelius rolled, his weapon now in his hand, but Flynt lunged with the point of his sword to pierce his wrist. He twisted the handle to give the blade greater purchase and skewer the hand to the ground.

Cornelius screamed like a cat in heat and reached for the sword to pull it free but a kick to his face sent him sprawling backwards, blood spraying from his mouth. Flynt stooped to extract the pistol from his grasp, tossed it to one side, then jerked the blade free, causing Cornelius to wail again. Cain moved to Ansel's side to ensure he was dead and then went off to retrieve the horses.

Cornelius began to squirm away, his heels digging at the dirt, his elbows working to propel him towards the bushes lining the track. Flynt allowed him to move a few feet before following, his steps slow, measured. Cornelius's movement became more frenzied and he began to weep.

'Tears, Cornelius?' Flynt asked. 'I wonder, did Madeleine weep before you murdered her?'

'Please,' Cornelius began, but could find no further words as he struggled to put distance between them.

'Did she plead for mercy?'

'I merely did what I was told to do…'

He had heard such protestations many times before. 'Ah, the defence of the weak-minded as a means of excusing their own barbarity.'

'Hawke told me…'

'I'll deal with him presently, but for now…' Flynt placed one boot on Cornelius's chest to bring him to a halt. 'I told you I would take pleasure in killing you, Cornelius.'

'Please, I didn't… I didn't…'

'But you did.' Flynt was calm. He was cool. He felt no pity. He placed the tip of his sword on the man's throat.

'I'll leave the island, you'll never see or hear from me again. Please… I have a family…'

That made Flynt pause. Of course, the man could be lying but Flynt was used to hearing men lie, both at the gambling table and in order to maintain their liberty, and he searched for sincerity in the man's voice and in his eyes. Not long before, back in London, he had killed a man in similar circumstances and though he, like Cornelius, deserved death, Flynt had been haunted by his own cold-hearted execution of it. He never knew if that man had a family and at the time he didn't stop to think. He never did. Until now.

Cornelius saw he had made a connection. 'A wife, two little ones, boy and a girl. They depend on me for their wellbeing.'

'Why should I believe you?'

'It's the God's honest, I swear it. They live in Nassau, down by the beach, near that shack we saw you in. In the name of the good Lord, Blake, I swear I speak true. Jenny and Michael, the little 'uns be called.

My wife is Abigail, she is Scottish, like you, but born here. You must believe me, and if you have no mercy for me, then have it for them. They need me.'

The level of detail was convincing. Flynt gritted his teeth, reached a decision and removed the tip of his blade from the man's throat, a bitter taste in his mouth caused either by the thought of depriving the family of a father or disappointment at failing Madeleine, even in death. 'I will make inquiry,' he said, his voice choked a little, 'and if I find you are lying to me then I will seek you out again and I'll finish the job.'

'Thank you, thank you, I won't let you down. I'll stop working for Mr Hawke, you'll see if I don't, and lead a better life. You've shown me the light, Mr Blake, that you have.'

Flynt turned away, the bile building in his gut. He had really wanted to see the man die. He still did.

He had walked a few paces when a scrambling sound behind him made him turn back. Cornelius had grabbed the discarded pistol and was bringing it to bear, a laugh growing. Flynt ducked but knew he wouldn't escape the ball and neither would he reach the man in time to run him through. Cornelius's laugh was cut short as a shot sounded and his chest erupted in blood, his arms thrown upwards as his body tumbled backwards.

Cain trotted up on the back of Ansel's horse, leading the other mount by the reins, smoke drifting from the pistol he held steady on the saddle. 'You know, Jonas, this continual saving of your hide grows rather tedious. Just once I wish you'd save mine, just to break the monotony.'

Flynt stared at the man's lifeless form, his anger growing. Cornelius had lied, that was certain. He had no wife. No children. But what grated on him was that he had convinced Flynt otherwise. He should have known better. He should have sensed it. But he hadn't and it had almost been the end of him.

'Why didn't you kill him?' Cain asked.

Almost shamefacedly, Flynt replied, 'An error of judgement.'

Cain held the reins of the horse to him. 'In our life sometimes a man just needs killing, without any thought to it. You think about it too much.'

Flynt didn't want to delve any deeper into the matter. 'When did you arrive on the island?'

'Early this morning. The lad Barbecue was desperate to reach Nassau. He arranged passage without my knowledge.'

So the boy had done what Hesikia had suggested. The quartermaster would be proud. 'I hope you managed to convince Cassie and Jonas to remain in Kingston.'

Cain gave him a blank look. 'What do you think?'

Flynt sighed. One day he hoped that Cassie would follow reason and not her own wishes. But then, if she did, she wouldn't be Cassie.

'How did you find me so quickly?'

'It wasn't difficult.' He waved a hand towards the corpses. 'Nassau is not so large, and you do tend to make an impression wherever you go, my friend. I had entered that tavern in search of some liquid refreshment when I saw those two individuals escorting you from the premises. I didn't think intercepting them among those cutthroats, or in the streets, was advisable, so waited for my moment.'

Flynt looked back the way they had come and then forward. 'We should move these bodies out of sight. Further into the bushes should do it.'

Cain grimaced. 'Awfully heavy work in this heat. Why not leave them for the birds to feast upon?'

Flynt was already hefting Cornelius by the shoulders. 'The birds can feast just as easily there, and it's best they not be found. They have a strict set of laws here and I was seen in their company.'

Cain dismounted and grabbed Cornelius by the feet.

'Where did you leave Cassie and Jonas?' Flynt asked as they returned for Ansel's corpse.

Groaning a little as he took the weight, he said, 'With Gideon and Mercy.' Predicting Flynt's next question, he said, 'They weren't hard to trace either. It seems you are all the talk of Nassau. For a man whose trade is subterfuge, you certainly draw attention to yourself.'

Having hidden the bodies a few feet beyond the road and covered them with twigs, loose earth and stones, they found themselves to be perspiring freely and somewhat out of breath. Cain leaned against the side of his new mount for a moment, both hands on the pommel but making no attempt to put foot to stirrup. 'Where to now?'

'Toby Hawke issued an invitation. Let's not disappoint him.'

The Paladin's bad seed had risen again.

35

Flynt followed the track leading to Hawke's house alone. He saw nobody nearby but he knew he was being watched, probably from the house itself. He was rewarded by the glint of sunlight on glass at one of the windows and he knew he was correct. The door opened as he neared and Hawke stepped out, closing a telescope, which he then handed to Bartholomew who loomed behind him like a bad memory.

'You wished to see me, Hawke?'

Hawke looked beyond him to the track leading to the road. 'Where are Ansel and Cornelius?'

'They had a subsequent engagement.'

'What form of subsequent engagement?'

'With death.'

Hawke's expression clouded. 'You killed them?' He smiled, his features even more resembling a death's head. 'Of course you did. It's what you do, is it not, Jonas Flynt?'

He waved a hand to Bartholomew, who emerged fully from the shadows with a musket pointed directly at Flynt. The move came as no surprise, though Hawke's use of his real name was a matter of concern. There was only one way he could know. As casually as he could he glanced to the left and right but saw nobody else, but he sensed they were there. Hawke was too confident for Bartholomew to be his entire force. He hoped Cain was on his mettle and continued to play for time. 'You are mistaken. My name is Joseph Blake.'

'Please, no more deceit. You are Jonas Flynt, a thief, a gambler and a killer. The use of an alias may not be unusual here but I feel sure Blackbeard will be most interested to learn of your connection to Gideon and will reassess his judgement. Not that you will live to see it, of course.'

He waved his hand again and Bartholomew stepped forward, raising the musket to his shoulder.

'I wouldn't do that, big fellow.' Cain stepped around the corner of the building, both pistols in his hands. Startled, Bartholomew glanced at him, giving Flynt the opportunity to draw Tact and Diplomacy.

'There are others,' he yelled, just as the barn doors flew open and three men burst out. A pistol shot rang out just as Flynt threw himself from the saddle to land painfully on his side and shoulder but he took aim between the horse's legs and dropped one of the men. The horse whinnied and galloped away.

Bartholomew swung the musket, but jerked to the side when Cain put a ball into his shoulder. Hawke darted back into the house just as Cain triggered his other pistol and Bartholomew staggered back against the wall, blood spurting from his chest. Flynt raised himself to a crouch as the nearest man charged towards him and triggered his weapon but the shot went wild. Flynt's aim was more deliberate. The man stopped dead in his tracks, his legs folding and collapsing underneath him. A shot exploded in the dust near to Flynt's feet and the third man was almost upon him. This one Flynt had seen before; he was the one Cassie had shot in Edinburgh. He dropped his spent pistols, snatched his cane from where he had thrust it into his belt and loosed the blade. The man showed more care than his friends for he had his pistol level and took aim. The musket barked from the direction of the house and he spun but remained on his feet so Flynt closed the gap between them and lunged, his sword piercing his chest. With a strangled cry, he pitched forward.

Cain lowered the musket. 'Seriously, Jonas, this relationship of ours is most one-sided.'

Flynt nodded his thanks, retrieved his pistols and began reloading as he climbed the steps. 'Hawke.'

Cain understood, threw the musket to one side, found his own pistols. 'What's the count?'

Flynt made a calculation based on his knowledge of the size of the building. 'Fifty, slowly.'

Cain nodded and charged his pistols as he moved back to the corner of the house.

'Don't kill him,' Flynt ordered. Cain rolled his eyes.

One, two, three...

Counting silently, Flynt flattened against the wall beside the open doorway and completed his reload.

Ten... eleven...

Around him, everything was still, the silence broken only by the staccato song of the mockingbird and the thrum of insects. He glanced at the bodies of the men lying in the dirt. None moved.

Fifteen... sixteen...

At his feet, Bartholomew lay propped against the side of the house, his head drooping onto his blood-soaked chest. Death all around, and Flynt was sickened by it. But one more death was necessary.

Sometimes you just have to a kill man, Cain had said.

Just to remind them who you are, Thatch had said.

Twenty... twenty-one...

He peeped cautiously around the door frame, but jerked his head back as a shot fired from the shadows within splintered the wood.

'I've got more than that for you, Jonas Flynt.' Hawke's voice reached him from inside. It was defiant but nerves cracked the veneer. 'And your friend, whoever that damned cur be.'

Twenty-five... twenty-six...

It was kind of him to alert them, Flynt thought. Cain would be at the rear by now and would have heard him, so would be doubly careful.

Thirty... thirty-one...

Flynt waited, hoping Cain's count was identical to his. It was a stratagem they employed when they worked the heaths, splitting up and converging on a coach from opposite ends, the count allowing them to hit two sides at once. It had worked then; it should work now. Unless the passage of time had slowed them down. Also, back then, the coach they intended to plunder was in plain sight, whereas Hawke was hidden somewhere in a structure the layout of which neither one had any knowledge.

Forty-two... forty-three...

'What are you waiting for, damn you?' Hawke grew restive. That was good. Nerves were not your friend in these situations. It took a cool head and a steady hand.

Forty-six... forty-seven...

Flynt closed his eyes momentarily and drew in a deep breath, then let it out slowly between pursed lips.

Forty-nine...

He pulled back the hammers.

Fifty...

As he spun himself into the house at a crouch, Cain crashed through the rear door. A shot rang out as Flynt pounded through the dim interior towards the sounds. Entering a kitchen area, Hawke whirled towards him, another pistol levelling but his trembling hand was too slow, giving Flynt the opportunity to fire first, the ball puffing blood at his shoulder, forcing him to drop the weapon. Cain darted through the open doorway, leaped across the room and clubbed him to the floor. To his credit, Hawke tried to reach for the dropped pistol but Flynt trapped his wrist under the sole of his boot.

'Did you see anyone else?' he asked Cain.

A jerk of the head to the yard behind the house. 'Just them, that's all. Did you expect more?'

Flynt glanced through a window to the rear, where the bedraggled group of men and women milled, looking with curiosity and a little fear towards the house. 'I did.' He returned his attention to Hawke. 'Where is he?'

Hawke winced in pain as he struggled to free his wrist. 'Where is who?'

Flynt pressed down harder. 'Don't play games. Where is he?'

Hawke cried out. 'Nassau! He's in Nassau.'

Flynt removed his foot and dragged Hawke upright.

'Who the hell are we talking about?' Cain asked.

'Daniel Hawke. You got here, so did he.'

Cain suddenly understood and raised his pistol. 'You have what you need, let's finish this old bastard and get after the young bastard.'

'No.'

Cain's teeth gritted. 'Damn it, Jonas, you're thinking too much again…'

'I have a better idea,' Flynt interrupted and pushed Hawke towards the door, then threw him down the steps and into the compound. He sprawled in the dust, moaning in pain, his fingers clutching his wounded shoulder.

'Take them,' he screamed at his servants as they clustered around him. 'Take them and kill them!'

None of them moved. They stood in a ragged half-circle, some looking at their master, the others at Flynt standing on the platform surrounding the house. A few of them carried work tools. 'He's yours, to do with as you please,' he said. 'Save him. Or not. It's up to you.'

'Jonas...'

Flynt held a hand up to silence Cain's protest. 'But while you make your decision, think of the way he has treated you and those who came before you. Think of the woman Mercy, whipped for no reason other than that he could. Think of the beatings you've had, the mistreatment, the lack of food, the lack of respect. Think of all that and then make your decision. And if you do what I believe you must, you can place the blame on us, or hide the body in the interior, but hide it well. Then perhaps make for the coast and find a ship to take you on their account.'

'Don't listen to him,' Hawke said, pulling himself to one knee, still clutching his bleeding shoulder. 'He doesn't know what life is like here. He doesn't understand. I've taken care of you, given you a roof, a home, work. All people need work.'

One of the men pulled himself erect and regarded Hawke with cold contempt.

'We ain't people,' he said, his voice frozen. 'We is property...'

He moved forward, the scythe he held swinging easily in his grasp. The others followed, silently closing in on Hawke, who tried to rise, but a painfully thin woman slammed one bare foot into his face. They clustered around him as he yelled orders and pleas and insults. And then the yells became screams.

Flynt jerked his head to Cain and they walked back through the house to retrieve the horses. They stopped on the front porch and Cain looked back.

'What if they leave him alive?'

A shriek rent the air. Long. Agonised.

'I don't think so,' said Flynt as he went in search of his appropriated horse.

36

They returned to Nassau at a brisk pace, Flynt taking the opportunity to apprise Cain of the events since they had parted company on the *Sprite*.

Cain laughed. 'As I keep saying, Jonas, you have an uncommon ability for drawing attention to yourself. I should give you lessons on how to come and go without being noticed. In our line of business such a talent is a necessity.'

There was considerable activity as they passed by the Mermaid. People milled around the entrance, clogging the stairway, and even in the dusty street. Madeleine's body had been discovered. He wanted to stop but he could add nothing. Anyway, he had business elsewhere.

They reached the harbour and found Barbecue standing on the somewhat ramshackle quay. He was sitting atop a wooden post to which some small boats were tied, talking to a slim, sallow youth who was slightly smaller than he and a few years his junior. Barbecue waved to them when he saw them alight from their mounts, climbed down from his vantage point and motioned the boy to follow him.

'Good to see you once more, Mr Flynt,' Barbecue said, prompting Flynt to flick an eye at the lad over the use of his real name. Of course, Barbecue had no idea that he'd used an alias. However, the boy displayed no reaction. 'This here's Israel, son to that Hesikia cove, what quartermasters for Blackbeard.'

'He doesn't like people calling him that,' Flynt warned.

Barbecue blanched a little, then said to the boy, 'I doesn't mean no disrespecting towards the captain, Israel.'

The boy shrugged, his eyes disinterested. 'Makes no never mind to me,' he said, his accent a curious amalgam of his father's and something more nasal, the likes of which Flynt had heard among those in Nassau who had either been born in these colonies or had spent a great deal of their life there. Madeleine McRobert had borne traces of it, too. The thought of her sent a shard of pain stabbing at his chest.

A relieved Barbecue returned his focus to Flynt. 'Mr Hands, he came by here a-looking for you. Recognised me straight off, he did, from when I had been shipmates with you on the *Sprite*. Does I know where the fellow is what joined his ship's company, says he, you mean Mr Flynt, says I, I don't know, sir, says I, which was the God's honest.'

'How did he react when you mentioned my real name?'

Barbecue frowned, for the first time suspecting all was not well. 'He didn't seem surprised none. Was I wrong in speaking it?'

Flynt waved the question away, for whatever damage it had done could not be undone. 'It's of no consequence. What did Hesikia want, Barbecue?'

Barbecue closed his eyes to recite from memory, as if repeating the words exactly was important to him. 'Tell Mr Flynt, says he, that I is desirous of speaking with him at his earliest convenience.'

A glance in Cain's direction revealed that his friend was suspicious, but he kept his counsel, no doubt because of the presence of Israel Hands.

'Where was I to meet with him?'

It was Hesikia's son who replied. 'My old da says he'll wait for you in the rum shack on the beach.'

Cain's eyes narrowed and he shook his head. Flynt's nerves tingled a warning.

'My compliments to your father,' he said to Israel, 'and I would appreciate it if he could wait for me a little longer. I shall be with him presently.'

The boy shrugged, giving no sign that he knew of any malicious intent, then without a further word turned and walked along the quay. When he was out of earshot, Cain remarked, 'You feel it too?' He sniffed the air. 'Smells like duplicity.'

Dan Hawke was somewhere on this island and he knew who Flynt really was. His father would have without doubt narrated the events regarding Gideon's trial and perhaps relayed a description of Flynt to him as someone who had interfered. The events at the farm bore that out. But had the younger Hawke taken such intelligence to Thatch and thence to Hesikia? That was the question, but Flynt agreed with Cain, there was treachery in the air.

He addressed Barbecue. 'How was Mr Hands when you spoke with him?'

'He didn't look too clever, and that's a fact.' He shuddered. 'Lost one of his pins, I sees. He seemed to hop around spright enough but the thought of my losing part of me like that fears me most deep and I ain't afraid to admit such.'

'His demeanour, Barbecue, when he spoke, what was it like?'

The young man's brow furrowed. 'He was weak, of course, but otherwise he seemed to be healthy enough to my eye.'

'Was he friendly?'

'He were, he were impressed when he clapped his peepers upon me, saying I had done a grand job in reaching here.' The young man preened. 'Said I'd make a fine addition to the crew, and to come see him once I has delivered the message to you.'

'I wouldn't do that,' Flynt said.

'Why not?'

Flynt thought of the imminent arrival of Woodes Rogers' fleet. 'The tide is turning for the Flying Gang. Joining them now is not advisable. Seek honest employment.' He knew his advice would fall on deaf ears, just as it had with Jack Sheppard back in London, just as it would have on him at their age, but he hadn't the time, or even the inclination, to press his case further. 'The ship you arrived on, when does it leave?'

Barbecue glanced at the position of the sun. 'The captain has transacted his business in Nassau and will sail right soon, I'd say.'

Flynt turned to Cain. 'You and Barbecue return to the ship and engage passage for us all. What's the name of the vessel?'

'The *Mary Jane*.' Barbecue pointed to a sloop anchored among the ships of the pirate fleet and various traders. 'She sails for the Virginia colony.'

All Flynt wished to do now was get off the island. They couldn't wait for Black John Corrigan's ship to set sail. 'Fine, give the master of the vessel anything he wishes but make sure he consents to take us.'

Barbecue was puzzled. 'What goes on?'

'Never mind, lad, just go with Mr Cain here and do as I ask.'

'I doesn't have no wish to go to Virginia.'

Flynt was already walking away. 'Then take Mr Cain to the master of the *Mary Jane* and then you can go your own way.'

Cain followed him. 'You're not thinking about meeting with this fellow, are you?'

Flynt shook his head. He was unsure whether to trust Hesikia, and that lack of certainty was sufficient to tell him that he shouldn't. 'I'll fetch my family.'

His family. They were his family. He was responsible for getting them to safety and he would do it, even if he had to kill every damned pirate on the island.

Cain flicked a finger to Barbecue, who reluctantly untied the rope tethering the rowing boat, then followed him down a treacherous-looking ladder. Flynt watched them man the oars and pull away into the channel, then veer towards where the *Mary Jane* sat at anchor. His focus then shifted a little to the left to find the *Queen Anne's Revenge*. Even from this remove, he could see activity on deck. Thatch was preparing to sail.

37

The shack was empty, the small cooking fire just outside the door still smouldered, and Gideon's pistol, returned to him after being cleared in the trial, lay on the sandy floor. He grew anxious. He'd asked them not to move from the shack or its immediate vicinity. Gideon would not have ventured into Nassau for there was nothing for him there, as Flynt had ensured they were well provisioned. He had Mercy back in his arms and he wouldn't leave her alone. Also, he would never have left without his pistol, not on the island. Thatch had said they were safe under his protection. He was not a man to make such a boast idly. That protection had been lifted.

He retrieved Gideon's pistol from where it lay, noting the tang of burned powder. He had managed a shot at least but the absence of blood meant the ball had not found its mark. He thrust the weapon into the back of his belt under his coat, then turned and ducked under the low threshold into the sunlight.

Hesikia waited for him, leaning on his crutches. He was still pale, his cheeks drawn, but he was alert and even smiling a little. Dicken and Hans stood on either side of the door levelling a pistol apiece at Flynt's head.

'Where are they, Hesikia?' Flynt asked.

'Safe,' the quartermaster said. 'We'll take you to them, if you wish. You was invited, after all.'

Flynt turned his gaze beyond the stern visage of the pirate to his right towards the drinking shack, which lay further down the beach, regretting that he had parted company with Cain.

'Dicken, Hans, take them barkers,' Hesikia said, 'for we wouldn't want them to go off accidental like.'

Two hands from either side reached out to pluck the pistols from Flynt's belt.

'That fancy cane, too,' Hesikia ordered. 'Mr Flynt here is something of a tricky boy with that sticker.'

The cane was removed from Flynt's fingers.

Flynt asked the question even though he already suspected the answer. 'Who else waits there, Hesikia?'

The smile returned. 'Someone who wishes a word or two. Someone you has wronged most grievous.'

Prodded into motion by Dicken to his left, they all trudged in the direction of the drinking shack. Hesikia made decent progress on the soft sand, having swiftly mastered the art of walking with the crutches. Dicken and Hans remained to the rear. No attention had been paid when he had been marched out of the Mermaid earlier and nobody on the beach took any notice. There were laws about theft and murder but paramount was the law that stated men should mind their own business wherever possible.

Inside the shack, Thatch waited at his usual table, the sunlight piercing the gaps in the wooden walls and slicing lines across the room while the area beyond the rough counter was in deep shadow. The bandolier criss-crossing his torso bristled with his customary three brace of pistols. But that wasn't what drew Flynt's immediate attention. Seated to Thatch's right, in a row of four chairs, were not only Gideon and Mercy, but also Cassie and young Jonas. They were all bound and gagged. Flynt had expected it but seeing them still jarred.

He made towards them but Hans stepped around him to block his path. Flynt noted that the pirate had the pistol he'd taken from him thrust into his belt. Dicken's hand landed heavily from behind on his shoulder to keep him in place. He would have his other pistol. It was useful to know these things. Hesikia hobbled to stand beside his captain, loyal as ever.

'If you've hurt them, Thatch...' Flynt snarled.

The captain laughed. 'No threats, Jonas Flynt, please spare us that, for you are in no position to deliver such. That is your name, correct? Not Joseph Blake, as you told me, but Jonas Flynt.'

Seeing no point in a denial, Flynt kept his silence.

Thatch pursed his lips then half turned his head towards the shadowed rear of the shack. 'This is the fellow, correct?'

Dan Hawke emerged from the gloom. 'Aye, that's him. Jonas Flynt, Gideon Flynt's son.'

Again, Flynt had expected him to be here, so easily maintained his impassive posture. He was tense, however. There was an atmosphere in this small space that was distinctly malevolent and it centred on the heavily bearded man seated at the table, staring at the goblet in his hands.

'You lied to me,' Thatch said.

Flynt made no response.

Thatch appeared unruffled by his silence. 'Understand this: it's not the lying to me about your name that is your crime, for I have many men who have chosen to hide their true self,' he said, his reasonable tone very troubling. 'It's the why of it that galls me. Trust, Jonas Flynt, that is what keeps this life of ours afloat. Trust in the man at your side, that when it comes right down to it, he will have your back.' He laid the goblet down, stood up and began to pace around the room. 'I don't give such trust easy, but I gave it to you, but all the while you were at the using of me.'

'If I had told you the truth would you have helped?'

Thatch considered for a moment. 'Perhaps not, but I would have preferred to have been given the opportunity. It pains me that I took you into my confidence, only to find that you were making sport of me, treating me like a chuckle-headed lobcock. In the short time I've known you I took you to be friend.' He stopped in front of Gideon. 'All you have said, all you have done, was to save this fellow's life, yes?'

'Yes.'

'And he is truly your father?'

'I told you that,' Hawke interjected, only to be silenced by a single look from Thatch.

'He is truly your father?' He repeated to Flynt.

'Yes.'

Thatch studied Gideon's features. 'I see no shadow of you in him.'

'I favour my mother.'

Thatch turned and peered through the gloom. 'Then she was a right homely woman, to be sure. Your features may be pleasing to certain ladies, but if it were a maid who bore them then she would be no object of desire.' Thatch moved on to Cassie, gripping her by the chin and forcing her head up. She struggled to jerk away from his grip. 'Now, here's a piece that would inflame passions. Got spirit, too.' He released her and she pulled away, anger blazing in her eyes. Thatch stepped in front of young Jonas, made a study of him, then of Cassie once more, then again of the boy. 'This here lad is this one's son, I'm told.' He made a show of examining him closer. 'But I do believe I see more there...' Thatch straightened and crossed the room to stand before Flynt. 'Aye, I see you in that boy, Jonas Flynt. He's your whelp, am I right?'

Flynt refused to answer, but his stomach churned. He felt the need to do something but he didn't know what. They were five to his one, including Hawke, and he had been stripped of all weapons. Gideon's spent pistol was still tucked under his coat but he would never have time to charge it. If ever he needed Cain to arrive unexpectedly, it was now.

Thatch sauntered back to the row of chairs. 'You gulled me. You claimed to be an emissary of the Fellowship but I would say you are not. You know of them, that be plain, but you're not a part of them, ain't that true, Daniel Hawke?'

'My father has had many dealings with them, as you know, Captain Thatch, on behalf of the Flying Gang,' Hawke said. 'He told me this man had nothing to do with them.'

'Your father didn't know everything,' Flynt said. 'The Fellowship had no faith in him.'

'That's a lie.'

'They didn't trust him.'

Hawke moved closer. 'Another lie!'

'He was cheating the Flying Gang and he was cheating the Fellowship.'

Hawke's anger rose. 'That's a damnable lie!'

He reached for a pistol but Thatch's voice brought him to a halt. 'There'll be no pistoleering, unless I be the one doing it, Daniel Hawke.'

Hawke's hand remained on the butt of his pistol but he didn't draw it from his belt. Even in the poor light of the shack, his anger-reddened face was pronounced. If Flynt could goad him into taking some form of action, that might give him the edge he needed to make a move. He didn't know what that move would be, though.

'I was sent here by the new Grand Master himself to deal with your father.' That, at least, may have been partially true.

Hawke sneered. 'You are no emissary of the Grand Master.'

Flynt forced a satisfied smile. 'His lordship would take issue with that. He and I are decidedly close. And he wished your father dead.'

His use of his lordship gave Daniel momentary pause, for only one who knew the Grand Master would know he was a nobleman, but then his grip on the pistol tightened. 'Damn it, another lie! You are nothing but a cheap street thief – I learned as much while being held in London by your friends.'

Thatch drew one of his weapons from his bandolier. 'This is my court and I won't be having you undermining my dutiful authority, boy. If you clear that barker, you'll never see sunlight again.'

Hawke glared at Flynt for a moment more, but then his hand dropped. 'I apologise, Captain, I forgot myself.'

Thatch kept his pistol level. 'Aye, that you did. Why don't you take yourself down to the dock, stand by the boat for my return. The chance is that your father will already be there waiting for us.'

He will be a long time waiting, Flynt thought, but said nothing, it being best he kept some distance from what had taken place on that farm. When the bodies were discovered he would have questions to answer, if he survived this encounter, but that was a problem for another time. He had learned to play the hand he had, and worry about the next one when it was dealt.

Hawke nodded his agreement and left, giving Flynt a sneer as he passed. One less opponent to deal with. For now, at least.

'So this is a trial?' Flynt asked Thatch. 'I thought the hearings were conducted at general council?'

'Hearings will be conducted wherever I say. And no general council can be convened, for some of the captains have taken their ships off from New Providence.'

Flynt recalled the activity he'd seen aboard the *Revenge*. 'I take it Woodes Rogers is closer than you thought?'

Thatch regarded him with contempt. 'That ain't no concern of yours. Back to the business in hand,' Thatch said, still holding his pistol. 'Most of all you have said and done since coming aboard my ship has been in order to save this man's skin. Lies. Deceit. Betrayal.' He moved behind the chairs. 'Such duplicity cannot go unpunished.'

Flynt made a move towards him but Hans punched him in the stomach then, as he doubled over, delivered a blow to the side of the head which sent him to the dirt floor, his mind spinning. He tried to rise but a kick to his ribs forced him down again.

'Gently now,' Thatch chided, 'I have need of him conscious.'

Dicken's strong hands hauled him back to his feet and held him steady as Thatch paced back and forward. Gideon strained against his bonds to look behind him but he was unable. Mercy's eyes were wide with terror. Cassie's anger still flickered, while young Jonas tried to remain brave, staring straight ahead, but the fear leaked through.

Thatch came to a halt midway between Mercy and Cassie but he ran his gaze over the heads of all four of his captives. 'Which one will it be? Spoiled for choice, I am. But I'll tell you what,' he addressed Flynt directly, 'I'll let you choose.'

He swallowed away the bile that had risen after the blow to the stomach. 'You're insane if you think I will do such a thing.'

A strange gleam beamed in the pirate's eyes. 'Insane? There are those that say I have a touch of the flap dragon, the Spanish gout, and that it has wormed its way into my brain.' He grinned. 'And who knows, mayhap it be true, for I've been with many a doxy and any one of them could have been poxed.' The smiled dropped. 'No matter, you must choose, Jonas Flynt. You must be punished for your breach of my trust. You will choose which death will bring you the least pain, otherwise I will decide and, mark me, that be something you would wish most intensely to avoid.'

Four pairs of eyes fixed on him, but Flynt couldn't discern what message they conveyed. He shook his head. 'I can't.'

Thatch began to parade behind them once more. 'Come now, of course you can. It's so simple, just speak the first name that comes to your head.'

'No name came to my head,' Flynt said, but he was lying.

Thatch gave him a sideways look, sensing prevarication, but letting it pass. 'I'll tell you what, then...' He motioned to Hans. 'Loosen them mufflers. I'll wager that each of them has a case to make as to why it should be them. They strike me as all being very noble and upright, even that Gideon one, Sam Bell as was, who was raider and murderer like the rest of us.' The men did as they were ordered. 'What say you, Gideon?'

'Choose me, Jonas,' Gideon said, his tongue sluggish from having the rag stuffed in his mouth.

'No,' Mercy said. 'Me.'

Cassie stared at Flynt, her eyes blazing, but he knew she wanted him to take her, if only to protect their son, who had straightened himself, his chin raised in defiance, but the glimmer of terror remained.

'Choose your pick,' said Thatch. 'It won't be difficult, for they all seem most eager to sacrifice themselves for your disloyalty.'

Flynt looked from one to the other, unwilling to say the name that had flashed in his mind. 'This is inhuman, Thatch.'

'Haven't you heard? I am monster. I have slaughtered men for no reason. This be nothing.' He waved his pistol towards the four figures before him. 'Reparations must be made. I told you I don't trust easy and you made of me a fool. Choose, Flynt, or I shall choose for you.'

'I won't choose, for you will not stop at one.'

'I give you my word. One is all I'll take.'

His heart sinking, Flynt was tempted to believe him, but still he couldn't bring himself to utter the name. He took a deep breath and shook his head. 'I won't play your game, Thatch. If you want to kill, you will do so, but mark this – if you pull that trigger, if you harm anyone here, I will see you dead.'

Thatch laughed. 'Boldly said. Now, speak a name, for I grow weary of this.'

Flynt was determined not to give the captain the satisfaction but his eye did momentarily fall on Gideon, who gave him a brief nod. Flynt couldn't bring himself to give voice but Thatch had already spotted the look between them.

'Aha, a choice is made!'

'I said nothing!' Flynt protested.

'You had no need.' Thatch raised the pistol. Gideon closed his eyes.

'Gideon,' Mercy breathed.

He opened them again and twisted to see her. 'It's my time, my love.' He jerked his arm against the bonds of his wrist as he tried to reach out to grip her hand, but the ropes were too tight. There was only an inch between their fingers but it might as well have been a mile. He gave her a weak smile and she began to weep.

'A very touching scene, what do you say, Quartermaster?'

By the way Hesikia looked from Hans to his right and then to Dicken behind Flynt, it was clear he was uncomfortable. 'Cap'n, a word...'

'No words, Hesikia. I am judge now and seek no counsel from you.'

'Cap'n...'

Thatch glared at him. 'Not another word!'

Hesikia fell silent but another glance at the seamen spoke volumes. Hans frowned. Behind him, Flynt was sure he heard Dicken curse under his breath.

'You don't have to do this, Thatch,' Flynt said, feeling he was standing on firmer ground now that the crewmen were visibly unhappy with their captain's actions.

Thatch thought about it. 'You know, you're right. I don't have to do this.'

Thatch looked around him, at his men, sensed the mood. Flynt felt his breath ease a little as he lowered the pistol.

'No, you're correct, I don't have to do this,' he repeated. 'Not when I can do this…'

He tossed the pistol from one hand to the other, raised it and fired directly into the back of young Jonas's head.

38

A moment's silence. A moment's shock. It seemed to stretch for an eternity until finally it was broken by Cassie screaming.

Jonas's head had snapped to the side. The chair had rocked on two legs. Settled again. His head drooped.

Gideon's eyes widened with horror.

Cassie screamed and struggled with her bonds but they held fast.

Mercy's mouth opened and closed as if gasping for air.

Cassie screamed and her jaw and teeth worked at the gag but it refused to dislodge.

Flynt's body chilled. Trembled.

Cassie screamed.

Thatch whipped the empty pistol against her head, again, then a third time, and her scream became a groan before she fell silent, unconsciousness perhaps a blessing. But she would wake and she would again experience the sudden loss of her son.

A roar grew in Flynt's throat then burst free as he lunged forward but a heavy blow from behind forced him to his knees again. He ignored the pain and the shriek in his ears to glare at Thatch. 'I'll see you dead.'

Thatch seemed bored as he slid another pistol from the belt across his torso. Alerted by the sound of gunfire, voices were raised outside and closed in on the shack. Thatch waved a hand towards the door. 'Keep them out of here, Dicken,' he ordered.

Dicken cast his eyes towards Hesikia, who nodded. The pirate left to assure the crowd that all was well, that the captain was administering justice, that they should all go about their business. Even though to Flynt's ear his voice was dull, lacked conviction, the inquisitive voices dimmed, quietened, nobody wishing to question Thatch's justice. Dicken remained outside, either because he felt he must or because he wished to have no part in the proceedings within.

Thatch studied Flynt once more. 'Now then, let's try again. Choose properly this time.'

He'd known the pirate captain was lying but those words still struck Flynt like a blow. 'You said you would take only one...'

Thatch was unconcerned. 'What can I say? I am mercurial. It must be that treatment for the clap they says I has, eh, Hesikia?'

The quartermaster swallowed, glanced at Flynt, an apology in his eyes. 'Cap'n, that lad wasn't doing no harm...'

Thatch glanced sharply at him. 'You question my actions?'

'Such matters should have been put to the general council...'

'Hang the general council,' Thatch snarled. 'I will hear no more of the fucking general council. It's dead, and goddamn it to hell. Hornigold's dream of a republic is dead and so will we be if we delay any further. This is my court. I am both plaintiff and judge. I was the one betrayed!'

Hesikia shifted his weight onto one single crutch. 'We was all betrayed, the whole crew what took this man here into their company. But that lad, he didn't do no betraying. None of these here folks have.'

Thatch pressed his fresh pistol against the back of Gideon's head. 'Not even this fellow?'

'Don't do it, Cap'n.' Hesikia dropped the crutch and clawed his own pistol free to aim it at his captain. 'He didn't betray you, he didn't lie to you, or to me, or to the men. This ain't correct, Cap'n, it ain't proper, and I'll ask you to lower that there barker.'

Thatch didn't lower his weapon; instead he swung it in Hesikia's direction. 'Damn you, Hesikia, don't question me...'

'It's my right and my duty, as quartermaster, to question. In an engagement your word is law, but we ain't in no engagement. Such actions go against the rules of the account, no summary punishments, no executions, not without vote straight and true by the men. You knows this.'

Thatch's voice lowered to a growl as he stepped around Gideon, his pistol unwavering. 'Are you challenging me, Hesikia?'

Hesikia swallowed again, the hand holding his pistol trembling. 'On this here issue, I am.'

Thatch advanced a few steps. 'I could kill you where you stand.'

'Then you would also have to kill my shipmate here, and Dicken outside, for they is witness and would report back to the ship's company what you did.'

Thatch glanced at Hans, who slowly raised his pistol and aimed it at him.

'Rules is rules, Cap'n,' Hesikia said, his voice gaining strength now that he had the upper hand. 'A ship has to have laws, you've said such yourself many a time, and them laws stand here on the dry as they does on the wet. We is thieves, something else you always said, not monsters. How many times has you had words with other captains over their cruelties? How they stain us all. Look to your own infamy, that of Blackbeard, stories of murder and torture that never happened and how that wounds you.'

Thatch wasn't for backing down. With his free hand he pointed at Flynt. 'Then this man goes unpunished?'

Hesikia jutted his chin towards young Jonas's body. 'Ain't that punishment enough? You took his son.'

For a moment Flynt thought Thatch would continue the debate, but with a small groan he lowered the pistol. 'I won't be forgetting this, Hesikia.'

'Nor will I, Cap'n, for this is a matter for the general council and I—'

Thatch's arm snapped up again and he shot Hesikia.

For a moment the quartermaster stared at him in shock, blood slowly seeping from the hole in his chest. Then he pitched forward.

Hans slowly, too slowly, overcame his surprise, but Thatch had already ducked away, his free hand snatching at another pistol. Hans cried out for Dicken just as a report cracked through the darkness and he staggered back, his heart pumping red onto the sandy floor. Flynt threw himself to the side, scrabbling to reach Gideon's pistol where it was tucked into his breeches beneath his coat, freeing it and swinging it level as he propped himself on one knee and sliding back the hammer. Thatch had pulled a further two weapons from his chest and they faced each other across the narrow divide, muzzles trained. The only difference was that Flynt's was not loaded.

Dicken burst in, his own pistol raised towards Flynt, but took in the scene and the two dead men.

'Easy,' Flynt urged. 'The captain here shot Hesikia and Hans.'

Thatch kept his aim fully on Flynt. 'He's a liar. You didn't do a good enough job in disarming him, for he had a pistol hidden away, as you can see.'

'How could I shoot two men with one pistol and then reload it, all before you returned?' Flynt said, his eyes on Thatch. 'Think about it.'

Dicken did think about it, and swivelled his aim towards Thatch. In return, Thatch adjusted the direction of the pistol in his right hand. 'Goddamn you for a mutinous bastard.'

'I heard Hesikia call you out, Cap'n, begging your pardon. But what he said was true and square. You has overstepped your bounds.'

Thatch snarled a little but kept his pistols trained on them both. The weapon Flynt wielded was useless, unless he could get close enough to use it as a club. He focused on the moment, resolutely keeping his eyes averted from his son's body slumped in the chair. He couldn't help Jonas, not now, nor did he have time for recriminations. That would come later.

'Well,' said Thatch, 'we have us here a nice little situation, do we not?'

'Give up your weapons, Thatch,' he said, his voice far calmer than he felt. His mouth was dry, his heart pounded and it was only by force of will that he prevented the muzzle from trembling. 'You'll get one shot off but you'll never get another. You get me, Dicken will take you down. You get him, then it will be my pleasure.'

'Shoot him, son,' Gideon said. Flynt ignored the order, but only because he had to. Had he a functioning weapon, he would have dropped the murderous bastard before now.

With sudden speed, Thatch dropped the muzzle of one pistol towards Cassie while alternating the aim of the other between Flynt and Dicken. 'I could do for her now, then take my chances that neither of you has aim decent to put one in me before I ducks away. I've seen Dicken shoot and let's just say he's most dextrous with a cutlass. What say you to that?'

Flynt allowed himself a humourless smile. 'You know I have both the speed and the eye.'

What he didn't have was ammunition. If he could somehow entice Dicken to hand him the pistol he'd taken from him, or reach Hans' body, he had a chance. But both eventualities seemed remote. To ask for the weapon would reveal his weakness to Thatch, while Hans lay too far away. He had to continue with the pretence that the pistol he held was primed.

'Then we have us a quandary,' Thatch said. 'I could end this girl…'

'And sign your own death warrant,' Flynt warned.

'Perhaps. But then, a life lived worrying about death is not much of a life, is it? That lad of yours perhaps understood that at the end.'

Flynt wanted to launch himself across the room to take the pirate captain by the throat, to squeeze and shake the breath from him, and it took every part of his self-control to hold himself back. That would kill Cassie, and he might never make it. He had to remain cool, focused, professional, in order to think of some way out of this stalemate. He didn't know if he could depend on Dicken. Whatever Flynt decided to do, he had to do it himself and hope that the Cornishman would somehow prove useful.

But what?

The answer came with something none of them expected.

The explosion outside was sharp, not terribly loud, but having the silence rent asunder in that manner, followed by cries of alarm, came as a surprise and they all started a little. Flynt recovered more quickly than the others and threw himself to the side just as Thatch fired, the ball missing him by inches. He rolled the two feet to Hesikia's prone form, snatched the pistol he'd dropped and fired in the direction of the pirate captain, but he'd dodged away to drop behind the counter.

Flynt held up a hand to Dicken, who appeared totally bemused by the speed of events, but who was sufficiently alert to draw a bead on the bar in case Thatch reappeared.

'My pistol,' Flynt demanded and to his credit the man immediately unhooked Diplomacy from his belt and threw it to him. 'Go and see what has occurred,' Flynt ordered and again Dicken did as he was told. Flynt plucked a knife from a scabbard at Hesikia's waist. The man stirred a little and groaned but he had no time to attend to him. Thatch was his priority. He backed away to where Gideon sat and sawed at his bonds, his eyes never leaving the makeshift bar. Flynt made a mental calculation as to how many pistols the captain had left in his bandolier. He calculated two, but he could even now be reloading. He handed Gideon the knife and motioned that he should free Mercy and the still-unconscious Cassie, then began to walk slowly towards the bar.

He was cool. He was calm. He had a task and he would complete it.

'Show yourself, Thatch,' he said. 'There's only one way out of this hovel and I'm between you and it.'

Thatch didn't reply. Flynt stepped closer. From outside came shouts and screams, but whatever had happened was of no concern to him in that moment. He advanced slowly, hearing movement from behind the counter, ready to fire at the first sight of Thatch.

'It's over,' he said. 'You lost.'

A rasp as wood shifted. Flynt came to a dead stop, weapon steady, breathing even, blood cold. When Thatch did not appear, he edged forward again, peering over the top of the counter, ready to jerk away, but Thatch was gone, a few loose boards in the wall shifting gently in the breeze. Flynt rounded the bar, got down on his knees, prised one back and scanned the beach beyond. It was only a few feet to the thick brush, where Thatch had managed to lose himself. He slammed the flat of his hand on the wall and cursed. He thought he had him but the bastard had slid away from under his nose. He turned back, the anger at losing Thatch diluted by sadness as he gazed at his family.

What was left of his family.

Gideon had cut Jonas free and gently eased him to the floor and Mercy was on her knees beside her grandson, ignoring the blood, holding his shattered head with both hands, keening softly while tenderly brushing his hair away from his forehead. Gideon crouched over her, his hand resting on her back, but his movements were awkward. Like Flynt he was uncomfortable with emotion. Cassie slumped in the chair, a trickle of blood sliding from the wound left by Thatch. Flynt moved to her side and gently touched her cheek with the back of his fingers.

'She lives,' Gideon said, his voice strangled as he struggled with his own grief.

Flynt still couldn't bring himself to look at the boy, not directly. He had failed him. He should never have allowed him to come on this trip. He should have insisted he remain at home, Cassie too. Or he should at least have thought of some stratagem to save him from Thatch. He had failed Jonas just as he had failed Rab Gow, the man the boy had called father. Just as he had failed Madeleine. Just as he had failed so many others.

He wanted to join Mercy and Gideon but he couldn't. He didn't believe he had the right. He blinked away the tears and swallowed back the bitter bile that burned at his throat.

Hesikia stirred again and Flynt knelt beside him. His stomach and chest were matted with the blood pumping from the hole left by the ball. His breath was ragged and he coughed, covering his chin with red. He looked up at Flynt, his eyes wide. He tried to speak but all that came out was a strangled groan and black, viscous liquid. He sucked in a sharp breath, which escaped very slowly, and then relaxed, his eyes staring but seeing nothing. Flynt laid him back down carefully and closed his eyes with the edge of his hand. He had liked the man, was glad that he had interceded when he did, but felt nothing over his death.

A fresh scream erupted from Cassie as she awoke and remembered. She fell on the floor beside Mercy, pulling at her son's corpse, dragging his head onto her lap. Again, Flynt should have knelt by her, should have held her close, comforted her, but he still couldn't move. He doubted if she would welcome his attentions anyway. He found himself unable to do anything. So he watched his family grieve, feeling helpless. No, worse – he felt useless.

And then Cassie looked up at him and reminded him that there was something he could do.

'Find him, Jonas,' she said, her tears streaking her face but her eyes bright with fury. 'Find him and kill him. It's what you do, so do it now.'

Flynt nodded, recovered Tact from Hans' body and left without a word, trying to understand why he felt an element of relief.

39

Dicken waited just beyond the door. Around him, men and women ran in all directions, but mostly towards the dock. When Flynt joined him, he said nothing but pointed out to the water, where a large ship, fully aflame, smoke billowing into the golden air of the setting sun, sailed towards three vessels at the mouth of the channel.

'The Royal Navy has arrived,' he said, his eyes not meeting Flynt's. 'They're blockading the passage.'

Giving the scene little more than a glance, without uttering a word Flynt walked briskly in the direction of the dock. Dicken followed.

'If I were to guess I'd say that there vessel's been torched to make it a fireship and set it on course for the navy lads, to cause confusion and terror. Ain't nothing more scarifying at sea than a fireship,' he said. 'And as it's a prize of Captain Vane's, I could also make me a serious calculation who it were what done it.'

Flynt quickened his pace. Dicken kept up.

'What happened back there, my mates and me had no part, you know that. The boy was innocent and there ain't no excuse for what Cap'n Thatch done, especial doing for the quartermaster and Hans.'

Flynt broke into a sprint. Dicken followed.

They dodged between bodies headed in the same direction, Flynt searching for sight of Thatch ahead of them. As they neared the dock, Dicken's pace faltered.

'We're too late,' he said, pointing towards a small boat being hauled towards the *Revenge*.

Flynt fancied he saw the burly figure of Edward Thatch at the prow, urging two men rowing to pull faster, pull harder, Dan Hawke looking back at them from the stern. He scanned the dock for another boat to commandeer and give chase, but knowing in his heart that they had too much of a lead and they would never catch them.

'Bastard marooned me here,' said Dicken, bitterly.

'Think yourself lucky that you still breathe,' Flynt said, his tone carrying little life.

Dicken expelled some phlegm, considered this, perhaps trying to calculate who would have caused his breathing to cease, Flynt or Thatch. Flynt wasn't sure himself.

'He should've done for you straight off, that were his mistake,' the pirate said as he squinted to watch the navy ships make haste to avoid the course of the fireship, then surveyed the remaining ships at anchor. 'Vane's ship has gone.' He looked back towards the open sea, his good eye honing in on one sail. 'Aye, there he goes, in the wake of the fireship. That bastard Thatch will do the same, I reckons.'

Flynt's focus was fixed on the small boat as it reached the *Revenge*, its anchor already being weighed, wishing he could somehow leap across the stretch of water between them and deliver justice for his son.

Dicken expelled another glob of slime. 'Damn that bastard Thatch for a-leaving me here. I hope he rots under the waves, I does, for he has done for me good and proper. I'll be gallows bait for dead sure.'

Flynt watched as the *Revenge* caught the wind and began to cut through the waves, his guts hollow. Dicken lingered for a few moments further then, realising that Flynt would neither speak to him nor try to take his life, headed off the dock towards town.

A stutter of cannon caused him to turn back. Smoke billowed from the navy guns but the balls fell short of Vane's little ship and Thatch's larger one. The fireship weaved dangerously on the currents so their position was not ideal, and firing upon the vessels was more symbolic than preventative. Flynt watched the useless engagement, less out of interest than the fear that this confluence of circumstance would mean he may never have the opportunity to bring Thatch to justice. His justice, not the government's, not the navy's, not even whatever remained of the pirate republic's. When he looked towards town again, Dicken was gone.

Flynt turned and walked slowly back to the shack.

Mercy keened over the loss of her grandson, Cassie still cradled his head in her lap, but her tears had dried and she looked up at Flynt as he entered with a question in her eyes. He knew what that question was and when he shook his head, he saw disappointment, perhaps even an accusation, but she said nothing as she looked down to stare at her dead son and whisper something as she stroked his blood-matted hair.

Gideon stood by Flynt's side and together the two men, both ill at ease, watched as the women they loved did their mourning for them.

40

When he finally made landfall the following day, Woodes Rogers was given a hero's welcome by not only the law-abiding citizens of the island but also a goodly number of the pirates, who sought to show their earnest wish to change their ways by cheering and waving. The people of New Providence, hundreds of them, lined up in two rows along the quayside and onto the beach, and as their new governor walked between them shouted huzzah for King George. A few muskets were set off into the air in celebration, causing some consternation among the squad of red-coated soldiers who acted as guard detail, but a weathered serjeant calmed his edgy men with a few well-chosen words.

Flynt and Cain watched the display from a distance, Cain leaning against the trunk of a tree and eating a banana, for which he confessed he had developed a taste. 'My God, the air is decidedly replete with the stench of hypocrisy. I'll be glad to be away from this god-rotting place.'

His tone lacked its usual jaunty air. He had taken the death of Jonas badly and viewed the island and its pirate inhabitants with a mix of cynicism and disdain. As Rogers made his progress between the ranks of celebrants towards the fort, he caught sight of Flynt and motioned that he should follow. 'I'm expected to make report,' he said to Cain.

He had already revealed to his old friend that he had another mission on the island. Cain straightened from his position and threw away the skin. 'I'll come with you.'

'There's no need.'

'There's every need. Word's already leaked out regarding your entanglement with Thatch and what caused it. I don't trust these people. When some of these men witness you in conference with this new governor they may decide to take some form of action.'

Flynt had already noted a few furtive glances in his direction and not all of them were friendly. The fact that Thatch had himself broken their rules didn't mean that they trusted him any the more.

'I don't need you to watch my back, Gabriel,' Flynt said.

Cain's tiny smile was melancholy. 'Oh Jonas, I think events, both recent and more distant, prove beyond any reasonable doubt that you really do.'

At the gates of the fort, Rogers was met by two men in clothing that had once been smart but years of island living had left them somewhat bedraggled. They declared themselves to be Chief Justice of the Island and Leader of the Council, and Rogers greeted them with all due solemnity, even though he would have been aware that in recent years both positions were somewhat redundant given the control of the pirate captains. He then produced a rather ornate scroll and began to read from it, outlining his credentials and his Commission from the King, and repeated the terms of the pardon for all the pirates present. Again, these were greeted with loud cheers, and some even sounded convincing. Flynt spied Benjamin Hornigold in the throng, waving his hat, and Captain England, who appeared to nod his agreement, but something in his expression told Flynt that he remained unconvinced by the government's blandishments. Dicken huddled with a handful of other men, no doubt discussing their options.

Rogers vanished into the fort, but first whispered to an aide and gestured in Flynt's direction. This, of course, prompted more looks from the erstwhile pirates. Flynt didn't care. He knew there to be men among them who had taken to the life out of necessity, or to escape the vicissitudes of honest life, and given his own past he was in no position to condemn them. However, there were those, if they were to be judged by their expressions as they regarded him, who sympathised with Thatch. Flynt had lied to them, deceived them and was now known to the man who had arrived to tear asunder their way of life. Flynt outstared them and though their eyes dropped away, he was minded of Cain's exhortation that his back must be protected. He was concerned, but not for his own safety.

'I need you to go to Cassie and my family,' he said.

Cain was also aware of the looks. 'And what of you?'

'I believe I will be safe with Captain Rogers. There's no appetite to challenge him and his forces yet.'

Cain slipped away as the aide, a foppish young man with a face pitted with pockmarks, waved a scented handkerchief in his direction and announced in a tone reserved for those who valued their own self-importance, 'The governor wishes to see you, sir, most immediate, if you please.'

Flynt inclined his head in acquiescence and followed the young man through the gates, ignoring the looks from those around him, though he was aware of Hornigold giving him a curious stare.

There being no accommodation yet ready for Rogers, he had elected to receive the more private welcomes and appreciation of the island dignitaries from a chair placed in the shade under the fort's battlements. Someone had brought him wine and fruit and he listened intently as the two men who had welcomed him loudly exhorted their delight at his coming. On seeing Flynt approach, he waved at the men and their acclamations.

'Forgive me, Mr Walker, Mr Taylor, but I would have a private word with this gentleman,' he said, prompting the two men to glare in Flynt's direction, clearly irritated that such a rough-looking fellow had been granted priority over them. Nevertheless they complied gracefully and Rogers waited until they had retired to a point where they could observe but not overhear.

'It's good to see you, Serjeant,' he said.

'And you, Captain.'

'Tell me quick, for I fear there be ears a-wiggling all around me. Who can I trust?'

Flynt made a small gesture towards Taylor and Walker, who Madeleine had pointed out to him days earlier. 'These were men of some importance in Nassau before the pirates came in force. I believe their welcome be heartfelt.'

'If somewhat effusive. You know them then?'

'We have never met but I heard talk from a local who I trusted.'

Rogers caught the past tense and the sadness in Flynt's tone. 'But you trust this person no more?'

'She was murdered.'

'By whom?'

Flynt told him quickly what he had learned of the pirates, the island, and its ways. Rogers listened, his only expression a hardness in his eyes when Flynt told him of the various deaths, omitting mention of Toby Hawke and young Jonas. The former out of a need not to incriminate himself, the latter because it was not a death he could relive.

'And what of that fellow Toby Hawke?'

Cain had ridden to the farm early that morning and reported back that all the bodies had disappeared, along with the slaves.

'He's vanished. His son also. He escaped with Thatch.'

Despite his best efforts, he couldn't keep a bitter edge from sharpening his words. Rogers noted it but didn't question. Flynt wondered if he had already been informed of events in that shack on the beach.

'And who was it who set the fireship upon us?'

'That was Charles Vane.'

'I knew that cutthroat would never see reason. So who among those who remain can I put my faith in?'

'Benjamin Hornigold,' Flynt said. 'He is a man of his word who has already taken the oath and will adhere to it. Others will follow.'

'And your advice on how to proceed? How do I keep order?'

'There was order here, of a sort, but I would recommend treading softly until you have need to stamp hard. Nassau was held as a republic…'

Rogers snorted in derision. 'Republics be nought more than a dream of those who have no power but desire it. They do not work in practice. We need a monarch, we need stability, we need a prince to look up to, to instil peace and goodwill…'

Flynt interrupted, history teaching him that, in general, kings and queens were interested more in their own peace and goodwill than that of their people. 'The republic here was successful, at least to a point, and what worked was down to Hornigold. Recruit him to your cause, Captain, and thus suggest some form of connectivity to the past.'

Rogers thought this over. 'I shall study upon it. My instinct is to come down most forcibly, to let these creatures know that His Majesty is lenient and loving to those who bend the knee, but vengeful to those who breach his faith.'

Without knowing it, Rogers might have been describing Thatch.

'I know men such as these, Serjeant,' Rogers continued. 'I know how they think, and they are merciless in their thievery. Some may turn away from that path, to be sure, but in others the lust for larceny is strong and they will return to it as sure as God is in His heaven. They must learn the folly of such a course.'

Flynt had no appetite to argue the issue, so let it go.

'And what of you, Serjeant? Your work here is done. You saved your father. You have provided me with intelligence. What will you do? Return to England?'

Flynt took a moment to reply.

—

Cassie was alone with Cain in the shack on the beach, the silence hanging between them suggesting that few words had been exchanged in the time they had been together. This was confirmed by the relief in Cain's expression as Flynt appeared in the doorway, and he motioned that they should talk outside.

'Gideon and Mercy have gone to confirm passage home,' Cain said. 'Gideon is armed and filled with rage. God help anyone who gets in his way.'

Flynt had told them about Black John Corrigan, whose ship was bottled up in the channel along with others. There was a possibility that he would be antagonistic, but Gideon would be a match for any former pirate who might bear ill will. Flynt didn't believe they would be molested, adjudging that men like the captain had their own future on their minds.

'How is Cassie?' Flynt asked.

Cain puffed out his cheeks. 'I never thought I would ever say this, but I believe she is broken. Cassie is the strongest woman I've ever met.' He corrected himself. 'No, the strongest *person* I have ever met. But that woman within yonder shack?' He twisted a little towards the doorway. When he looked back Flynt swore he saw tears glistening in his eyes. 'That's not the Cassie I know, nor less the one you ever knew. It wasn't only young Jonas who died yesterday, I fear.' His sigh was ragged. 'I think we've all died a little.'

Flynt made for the door, but stopped when Cain placed a hand on his arm. 'Tread softly, my friend. She blames herself.'

Flynt nodded his thanks and ducked under the doorway. Cassie had not moved from the three-legged stool on which she sat. She stared at the sandy floor, her eyes dull, her arms loose, her hands cupped limply on her lap.

'Cassie,' he said, gently.

The faint tick of a nerve in her face told him she'd heard him. He steeled himself and moved closer, getting down on his knees, his face level with hers. 'Look at me, Cassie.'

Her eyes rose, very slowly, and he saw her pain clearly. He loved those eyes – he'd loved her from the first moment he'd seen her, when Gideon brought Mercy and she home for the first time. She had been a gangly lass, all legs and hair and a smile that was as bright as the Caribbean sun, and Flynt the boy fell for her, though he didn't know it. Seeing them now so haunted struck at him like a sword thrust.

'It wasn't your fault,' he said.

Tears welled in those eyes.

'It was mine,' he said, his voice breaking. 'I should have done more to prevent you and he from coming. I should have…'

She moved then, her hands rising, her finger softly touching his lips to silence him, her head shaking and in so doing dislodging tears that had dammed in the corners of her eyes.

'No,' she said. 'The fault is mine. You urged us not to come and you were right. You knew it was dangerous but I was too stubborn, too thrawn, to listen. I put my son… our son… at risk, not you.'

Flynt felt his own tears forming. It had been a long time since she had spoken so tenderly to him and it took tragedy to bring it about. He knew, however, that the root of that tragedy lay in his own actions, years before, and for that he would forever blame himself.

'I cannot comprehend that I will never again see him or embrace him or even berate him for tardiness or some minor infraction,' she said. 'The loss of Rab was difficult to bear but this… this… is too much.' And then her arms were around him. 'Too much, Jonas, too much…'

He held her close, no words coming to him that would adequately assuage her grief. And in that moment he knew that he would never, ever break the hold that she had over him. He cared deeply for Belle, but his connection to Cassie was even stronger though they could never be together, no matter how close they were in that moment. He had failed her too often. He had brought too much death into her life. So he held her and she held him, and finally Jonas Flynt allowed the emotion to flow and together they wept for the son he had never come to know.

41

Woodes Rogers paid no heed to Flynt's advice to tread softly. He almost immediately declared a state of martial law on the island and appointed one of his staff to organise a survey of all cargo held in those ships confined to Nassau harbour. Rogers also decreed that island justice would be placed under the jurisdiction of the navy, and so wished an Admiralty Court to be formed with officials of his own choosing. All of this caused alarm among the former pirates, for some holds housed goods seized by force, and discovery would lead to the rope or lash, just as Vane had predicted. It was further revealed that the new island administration expected them to undertake gainful employment by improving the facilities and structures and creating new ones. Roads were to be improved, ground cleared, buildings repaired or demolished or built. Settlers were offered plots of land with the proviso that they be developed within one year. Pirates were, by their very nature, free spirits, if considerably larcenous and even murderous, and the notion of being part of civilisation and adhering to its associated strictures was anathema to them. Even the provision of supplies, including alcohol, was insufficient to appease their wilful natures, and so they began to desert the colony and return to their old life. In the dead of night, ships cut their cables and slipped away.

Benjamin Hornigold was enlisted to bring those raiders operating in the territory to justice, his experience as pirate deemed invaluable as hunter, causing Flynt to consider whether the notion of using a thief to catch a thief was something Woodes Rogers had been taught by Colonel Charters. After all, that was the very foundation of the Company of Rogues.

Flynt regularly conversed with the new governor, advising him where he could, informing him of what he knew of the island's life. Of Toby Hawke nothing was found. Even the bodies of the men Flynt and Cain had killed were gone. He learned that three slaves only were found at the farm, and they were taken by another landowner. Flynt wished they had joined the others, wherever they went, but they were too old or infirm for a life on the run. He hoped their new owners treated them kindly.

All of this came to pass while Flynt's family waited for Captain Corrigan's vessel to be granted leave to sail. They lived in the small shack, while Flynt and Cain took accommodation in the Mermaid. The remaining pirates gave him a wide berth, which was all to the good for he had little appetite for any further encounters. What he did know was that Mary Bowers had assumed ownership of the establishment, she being the only one among the girls who had her letters and could perform addition without the need of fingers and toes. He had no clue whether that would have been what Madeleine wished, but the arrangement did remind him of Belle St Clair and another Mary, Mother Grady. He had been away from London for a few months now and he knew not whether the tough old madam still breathed. He often thought of Belle, most fondly but not without some measure of sadness, for he felt he had lost her from his life forever. Cassie remained in his life but, their moment of tenderness apart, there was nothing there but friendship and perhaps a sense of loss for what might have been. And grief, of course; there would always be grief.

He and Cain laid claim to a table in the corner of the tavern with a clear view of front and back. The Mermaid still managed decent business despite the reduced number of men, who continued to couple with the girls in full view of the patrons or more discreetly in a room above or in one of the alcoves. Cain had met an Irish lass named Anne Bonny, whose state of marriage was of no inconvenience to him or to her. Her husband was away at sea and she saw no reason to remain faithful. She was most comely but had a temper on her that would rival even that of Thatch. She also had a wanton streak, evidenced not just by her willingness to engage in an extramarital fling with Cain, but she also at one time, while much the worse for brandy to be sure, suggested that Flynt join them in their bed. He might also bring with him one of the girls of his choosing. Flynt politely declined the offer.

When not in the Mermaid, they walked the island, sometimes with Cassie but more often not, as she had taken to spending much of her time in solitude, which brought Flynt pain, for there was nothing he could do to alleviate hers.

Cain shared at least part of the ache Flynt felt. 'She will get through it, Jonas,' he said as they watched the tavern commerce.

'I know,' Flynt said.

Cain paused. 'So will you.'

Of that Flynt was not so sure. He continued to carry the guilt and he knew he would forever bear it. It would grow lighter on occasion, and perhaps there would be days he would forget it, but it would always return, along with all the other guilt and regrets he bore. That was his life.

Cain was smoking a pipe and he tapped out the used tobacco on the tabletop. 'You will not be returning with us, will you?'

When Woodes Rogers had asked him what he intended to do, he had been at first unsure, but then the answer came and it surprised him.

'There is nothing for me in London,' he now told Cain.

'And there is something for you here?'

'Thatch is still out there. As is Dan Hawke. And I won't leave these waters until I have found him.'

'Revenge is a bloody business.'

'Blood is my business,' Flynt said, flatly. 'Yours too.'

Cain could do nothing but accept that. 'And if you find him, what then?'

Flynt stared across the room at nothing in particular, perhaps searching for the future. 'Who knows? A fresh start, perhaps. There are only memories at home and few of them good.'

'And Cassie?'

Flynt shook his head. 'She is better off without me. There are too many shadows in my life and Cassie has experienced enough darkness.'

'Perhaps you should give her the choice.'

'She doesn't belong here, or anywhere in the colonies. She is better with her family.'

Or what remains of it, he thought.

Woodes Rogers was sick, as were a number of the soldiers, seamen and settlers he had brought with him. The epidemic had raged through the newcomers, and some had already died, with more about to succumb. Rogers claimed it was the taint of rotting flesh and hides in Nassau, coupled with the tropical heat. At Rogers' urging Flynt kept his distance from him, a scented kerchief across his mouth.

'I can't have you stricken with this infernal flux, Serjeant,' he said, his voice hoarse, as he reclined on a cot in a newly refurbished room within the fort. He had lost no time in restoring the fortifications and installing ordnance, for the Spanish were beginning to agitate once more and there was a fresh threat that they would make a bid to retake the island now that the pirates had left, believing it to be vulnerable. Rogers had every intention of proving them wrong. The walls had been strengthened, the garrison housed in newly built shacks with roofs of palmetto leaves and a civilian militia formed to defend the town against any attack, whether by the Spanish or Charles Vane, who had signalled his displeasure at being chased away.

A naval surgeon fluttered around the governor and was waved away irritably. The man sighed and retired, rolling his eyes towards Flynt. 'The governor is a most difficult patient,' he muttered as he passed.

'I heard that,' Rogers said.

The surgeon was not concerned. 'You were supposed to, sir,' he said, then closed the door behind him.

Rogers grinned but it was wiped away by a cough, which he stifled with his own kerchief. 'He's a good man, but he does fuss.'

'You wished to see me,' Flynt said, for in truth he had no desire to remain within these walls any longer than he had to. He had already been kept waiting for two hours as a convoy of officers and local politicians were ushered in and out of the room.

Rogers laughed a little but again a cough racked him. 'My God, man, I am decidedly unwell, can you not temper your natural tendency towards bluntness this once? Nate Charters told me you were a forthright man, bordering on insubordination.'

'I have little time for social niceties and I've already been cooling my heels in the courtyard for far too long.'

'You have somewhere else to be?'

Flynt sighed. Charters had clearly trained Rogers too well. 'Tell me what you need me to do and I'll do it, if I can.'

Another laugh. 'The wait will be worth it, you have my word on that, for what I have to tell you will please you greatly.'

Something in his tone made Flynt temper the truculence he normally adopted when dealing with a commander. He waited for Rogers to amplify but he had erupted into a paroxysm of coughing behind his linen. When he took it away there was no trace of any blood, which Flynt took to be a good sign.

Rogers took a moment to catch his breath. 'I have this day received news from the British-American colonies.' He leaned over to expectorate into a bucket at the side of his small bed. Again, Flynt detected no sign of blood. Rogers sat back and closed his eyes, suddenly very weary, so Flynt waited until he spoke again. The eyes opened. 'The man Blackbeard has gone to ground near to Charleston.'

Flynt felt his blood thrill but he said nothing.

'He has been given safe haven by the governor there, who seeks to enrich himself by aligning with that damned cutthroat. The populace are most perturbed by this development. From the safety of the Outer Banks, Thatch enjoys the freedom to harry shipping at will and this has displeased me hugely. I've been in correspondence with Lieutenant-Governor Spotswood of Virginia, who has petitioned for assistance in dealing with the threat once and forever. I was vexed that the man slipped through my fingers when I first arrived, so I have arranged passage for you to the mouth of the James River, where you will liaise with the captains of two Royal Navy sloops, the *Pearl* and the *Lyme*. I've had my master of the harbour clear the backlog of vessels at anchor and you will leave within the hour. My secretary will provide you with the necessary bona fides.'

The excitement of finally being given the opportunity to face the man one final time surging through him, Flynt gave the governor a short bow and turned to the door.

'Oh – and Flynt?' Rogers called after him. Flynt turned to face him. 'I'm sending you for a reason. You have a personal stake in this hunt, yes?'

'Correct.'

Rogers nodded. They had never discussed this matter, but Flynt was not surprised that he knew. 'I appreciate that your passions will be high, but understand this: I cannot have this particular viper at liberty to mock me further. Do not let your hatred for him blind you to the task at hand. An angry man can be a careless one. Find Edward Thatch, and take him into custody to face the King's justice.' He paused. 'Or bring me back his head.'

42

Once again Gideon's shack was empty, but this time it was different. Where before belongings had been left behind, this time nothing of a personal nature remained. As soon as Flynt saw this it struck him – Woodes Rogers had said he was expediting the clearing of the harbour. He stepped onto the beach and looked at the masts punctuating the blue waters, some already underway, unsure which vessel was the one on which they had passage.

'They had to depart, Mr Flynt.'

Barbecue approached him from the bush behind the shack. Hesikia's son Israel was with him and he regarded Flynt with a surly expression, as if blaming him for his father's death.

'The lady, Miss Cassie, she asks us to deliver you a message.'

Flynt almost screamed. While he had been loitering within the fort awaiting the governor's pleasure, they had received word that their ship was weighing anchor. 'What was the message?'

'She says "Tell Jonas," begging your pardon for being so familiar, Mr Flynt, but this is what she says to me, "Tell Jonas that it is my turn to leave without a goodbye."'

'How long have they been gone?'

Barbecue rubbed his chin and wrinkled his face. 'Hard to rightly say, doesn't seem overlong, do it, Israel?'

Israel made a show of calculation. 'I doesn't know correct, but maybes fifteen minutes...'

There was perhaps still time. Flynt raced across the beach towards the harbour, his feet sinking in the soft, golden sand, ignoring the looks from those lounging on the beach, fewer now than before. He knew Barbecue and Israel were following him but he paid them no heed. He pounded onto the wooden dock, scanning the small boats lined up at the moorings taking on cargo and passengers, hoping to see them but feeling disappointment grow. He reached the end of the quay and stared out at the channel. He could take one of the boats, have the oarsman row him out, but he didn't know which ship they were on. He cursed himself for not walking out of the fort when he saw that Woodes Rogers was playing power games. He cursed himself for not asking Gideon the name of the ship.

'It's the three-masted brig pulling away now.'

Cain's voice startled him. He turned to see his friend bearing a sad smile. 'You didn't leave with them?'

'I would say that query is somewhat irrelevant, given I stand here before you.'

'Why did you not go with them?'

'And leave you to enjoy all the entertainment this part of the world offers? And who will save you when you inevitably find yourself in peril? Besides, I don't think Mistress Bonny is quite done debauching me.'

Flynt, despite his continuing disappointment, smiled back at him, then pulled his spyglass from his pocket to fix it on the ship. He scanned the decks, hoping for sight of his family. Of Cassie.

'I can't see her.'

'She'll be there,' Cain said.

'I promised her I'd say goodbye.'

'She understood. It all happened too swiftly. Barbecue was sent to find you but you were by that time closeted with Woodes Rogers.'

And then he saw her, emerging from below decks and moving to the rail, her gaze directed towards him. He waved. He had never waved before but he did now and it felt natural.

'She doesn't see me,' he said.

'Wave again, it's most becoming.'

Flynt gave him a swift, dry backward glance, but nonetheless waved again, this time removing his hat and sweeping it back and forth over his head. This brought results. Cassie had been leaning on the rail and now she straightened, tilted her head a little to the left and her features softened with a smile. She hadn't smiled at him for such a long time, and he wished they had not been separated by an expanse of water. She waved back. Behind her, Gideon and Mercy had also appeared topsides and joined in. They were smiling. They were weeping. Tears burned at his eyes.

Flynt watched as the ship headed beyond the mouth of the channel and out to sea, its outline sharp against the sun's reflection. His family had long since vanished from his sight but he kept his eyes on the sails until they rounded the point of Hog Island. Only then did he turn away and return to Barbecue and Israel at the end of the dock. He was used to leaving people behind; he was unused to being left behind. But he had an objective. He had wished to tell Cassie that he would soon confront Thatch and had wanted to assure her that he would pay for her son's death. His son's death. Their son's death.

By the time he reached the boys he had regained his cool demeanour, at least externally.

'I'm leaving the island, Barbecue,' he said.

'We is too, Mr Flynt, me and Israel here. There ain't nothing for us in Nassau, not now.'

'Where will you go?'

He ensured nobody was close by. 'We has heard that Captain England be shipping out come nightfall, sailing for the Africas where there is rich pickings for an enterprising crew.'

'You still wish to go pirating?'

Barbecue's grin was wide and open. 'Adventure, Mr Flynt, is what we wishes, and that ain't available to us boys unless we submits ourselves to the navy's mercies. And merchantmen ain't much better.'

Flynt considered trying to talk them out of that path but decided against it. It was not his business and Barbecue was too like Jack Sheppard in London. He would choose his own way and that was that.

'Good luck to you, Barbecue,' he said, holding out his hand. 'My thanks for the assistance you have shown me and my family.'

'It has been a proper pleasure, Mr Flynt, and here's my hand upon it.' They clasped hands and then released. 'And if you ever wishes to go a-pirating yourself, then you must seek me out. It would be an honour to sail with you. You would make a fine captain, and there ain't no mistaking that.'

'An unlikely occurrence, given I am no seaman.'

'There's more to being master of a ship on the account than navigating and knowing about sails and such. There's coves what does all that for you. But a captain must be able to lead his men, to have their faith, and it be my reckoning that is something you is most capable of.' He smiled again. 'Cap'n Flynt. Has a ring to it, don't it?'

He'd been called captain before, but from the tongue of this young lad of the salt, it did sound different. Comfortable, even.

'If ever I do decide to take to the sea, I will find you.' The boys began to walk back towards the town, but Flynt called out after them. 'Barbecue, what is your given name?'

The boy turned, but kept walking backwards. 'Don't often have much use for it,' he shouted. 'But if you wish to know, it's Silver, John Silver...'

Part Four

North Carolina's outer banks

43

The naval lieutenant reading the letter from Woodes Rogers was suspicious. He was tall and carried an imperious manner, one perhaps instilled in him by his years with the Royal Navy, but it might just as easily have been bred into him.

'I am not prepared to berth civilians on this expedition,' he said, displaying open irritation at having two individuals foisted upon him who he no doubt believed would prove to be little more than ballast.

'We're not civilians,' Flynt said. 'As our bona fides inform you, we are agents of the governor of New Providence island…'

A disdainful curl of his top lip conveyed what Lieutenant Robert Maynard thought of that title. 'Nonetheless, this is a military matter and I can't have my attention diverted to the care of two minor government functionaries.'

Cain bristled at that, but Flynt maintained a calm demeanour. He had dealt with men like this before. 'We're more than capable of looking to ourselves, Lieutenant.'

Maynard studied them more closely now, taking in their black apparel, but more importantly their weapons.

Flynt tapped the papers still in the officer's grasp. 'As you will see, there is an addendum by Mr Spotswood, the Virginia lieutenant-governor, at whose instigation this action is being taken. He instructs you to afford us every assistance, just as we have been instructed to assist you in this endeavour.' Maynard slipped the second document on top of Rogers' letter. 'You may also notice,' Flynt said, 'that it is countersigned by your own captain of the *Pearl*.'

Maynard, in command of two sloops, the *Ranger* and the *Jane*, had sailed from the James River before they arrived in Virginia, but his commanding officer had sent them on another small vessel to catch him. The signatures on the orders did little to alleviate his annoyance, though he was bound to obey. He allowed a long, harsh breath to escape him as he folded the documents and handed them back to Flynt with a flick of the wrist, then adjusted the rolled-up sea chart tucked under his arm.

'You will bunk with the men or sleep here on the deck,' he said, 'for we have no amenities to offer.'

Flynt looked around them as he folded the papers before thrusting them in the pocket of his coat. They were anchored in the mouth of a river, which opened up to a wider expanse of water. Heavily wooded land stood on both sides growing from what Flynt thought might be wetlands. It was swamp and not somewhere he would choose to take a stroll. 'And where exactly are we?'

Maynard took a moment during which he clearly contemplated not parting with the information. 'How familiar are you with these waters, sir?'

'Obviously not familiar at all.'

That vexed the officer even further, but he unravelled the chart he carried, folding it to reveal one section and holding it out so they could each see. 'We are for the moment in the mouth of the Alligator River, feeding into the Albemarle Sound, do you see?' His finger traced a line south around the coast. 'Thatch has anchored his sloop the *Adventure* in Pamlico Sound, close to Ocracoke Island, there.'

'The *Adventure*?' Flynt asked. 'What happened to the *Queen Anne's Revenge*?'

'He ran it aground, divested himself of some of his crew, we believe in order to increase his own share of the booty.'

Without Hesikia's calming influence, Thatch had turned pirate on his own men.

'Does he know we are coming?' Cain asked.

'We have kept our plans most secret, but there is always the chance that someone will have alerted him. The administration of North Carolina is not as virtuous as it might be.'

'Then he could have upped anchor and fled by now?'

'That is not our intelligence.'

In Flynt's experience, such intelligence was often not as intelligent as it should be. Cain had been studying the deck of the *Jane*, then squinted across the water to the *Ranger*. 'You have few guns, I see. Some swivels only.'

'Small arms certainly, but with good English guts behind them sufficient to the task.' Flynt caught a note in Maynard's voice that suggested he also harboured doubts about the lack of ordnance, but he would never voice criticism. He was an officer in the Royal Navy and he had his orders. 'Thatch's crew has diminished considerable and our intelligences tell us that he also lacks heavy ordnance. I have sixty men under my command, here and on the *Ranger*, all hardened seamen,' Maynard continued. 'We are more than a match for a raggle-taggle band of miscreants without doubt befuddled by spirits much of the time.'

'Why these small ships? Why not the *Pearl*?'

'The *Pearl* carries too much draught to venture into those waters.'

'And how far is this island?'

Maynard rolled the chart again. 'In a few minutes we shall weigh anchor and sail around that headland, through the Croatan Sound and strike at Thatch before he knows it. This rogue has been active too long, gentlemen, and it is the Royal Navy's intention, and mine, to end his terror once and for all. This will be done efficiently and at speed, on that you have my solemn oath.'

Whatever the source of the navy's intelligence, it proved accurate, for Thatch remained at anchor near to the island, nestling just off Ocracoke, which to Flynt's eye was little more than a low-lying streak of land situated at the mouth of a vast lagoon, and a much smaller neighbouring island.

They arrived at dusk and anchored to wait until morning to attack. They were so close that they could hear the sound of carousing floating over the surface of the water to them, and Flynt was certain the loudest voice of all was Thatch, who had plainly rejected all the rules to which he had appeared to hold dear. He warned Maynard that a drunken Thatch was the most dangerous of all, but the lieutenant dismissed him.

'Our intelligence is that Blackbeard has little more than twenty crew remaining. We outnumber them three to one.' Maynard regarded him with cold superiority. 'We shall prevail, for we have right on our side.'

Flynt knew that arguing with the man would bring no benefit, so he walked back to Cain, who leaned against the rail. 'What did he say?'

'We have right on our side.'

Cain was unimpressed. 'A broadside can wither even the most righteous.' He studied the tall naval officer. 'Does he know what he's doing, do you think?'

Flynt considered his words. 'From what I've seen, I think he is a most competent man, though a little arrogant, which they hand out along with his rank. He's been given orders and he will follow them to the letter.'

'Even when they have not equipped him properly for the task?'

'Even then.'

'So what do we do?'

Flynt exhaled harshly. 'Trust in luck but keep our powder dry…'

44

Nobody slept well that night. Flynt and Cain checked and rechecked their weapons while around them iron scraped on whetstone as the men sharpened cutlasses. Had they not been so close to the enemy, Maynard would even have had the gun crews run drills to ensure their speed and accuracy was at an optimum. The mood was sombre, for each man had no idea what the morrow would bring. Those who could write scribbled words on paper and assisted those who could not to compose thoughts to be sent to a loved one. Even Cain was subdued. He, like Flynt, believed the quiet before a battle was the worst part. When hostilities commence there is much to do, staying alive being uppermost, but during the waiting a man's mind turns to his own mortality.

Dawn broke to reveal a thick mist enveloping them and blocking the sun. The island to their starboard was a dark line, the smaller one not visible at all. Maynard ordered the anchor to be hauled up and, catching what breeze there was, inched forward, sharp eyes alert for sight of the inlet. The small island hove into view and Maynard had the *Jane* wheel to starboard into the strait that separated it from the larger, the *Ranger* following suit. No sound emanated from the pirate ship, which was still not visible, so the crew had perhaps drunk themselves into oblivion. Thatch would have watches on duty, though; he was too cautious not to, even if he felt secure.

Finally, the spectral lines of the *Adventure*, anchored close to the western edge of the island, began to form. Flynt and Cain each drew their pistols, even though at that range they were useless, but it felt good to have them to hand. The gun crews had already prepared what little ordnance they had and waited patiently for the order to fire. Maynard stood on the quarterdeck, his eye to his spyglass, occasionally issuing instructions in a low, but calm, voice. He appeared relaxed, confident, and Flynt wondered how he had spent the night. Did he sleep well, content in the surety that they had right on their side? Or was it fitful, restless, anxiety over the impending action murdering sleep?

The first they knew of the sandbank was a rasping under the keel which grew to a loud grating noise and within seconds both sloops had ground to a halt. A momentary flash of annoyance crossed Maynard's face as he glared at the local pilot who had been guiding him, before he resumed his previous equanimity.

'Let's lose some ballast, Mr Willard,' he whispered calmly to his first mate. 'Softly now, for we have not yet been observed.'

The mate saluted and scurried off to do as he was told. Men scampered down companionways and reappeared carrying heavy stones and lumps of iron, which they carefully lowered over the port side in order to reduce the sound of the splash. The midshipman in command of the *Ranger* had his men do the same. Maynard maintained his observation of the mist-shrouded *Adventure* as if this was no setback at all. Flynt admired his coolness.

They had hoped for the element of surprise, but their ignominious grounding gave Thatch's men the opportunity to raise the alarm. By the time both sloops had risen in the water and cleared the sandbar, there was activity apparent on the *Adventure*'s deck, though the men involved were little more than ghosts in the mist. The *Ranger* freed itself first and proceeded to enter the sound. As the *Jane* floated free, a familiar voice found them.

'Damn you for villains, who are you and from whence have you come?'

'Thatch,' Flynt muttered.

'And so it begins,' Cain said.

Maynard calmly ordered the British ensign be run up the mast. Once it fluttered in the slight breeze, the bright red flag discernible even in the mist, he raised his own voice in reply. 'As you can see, we are no pirates.'

There was a silence for a moment, during which Thatch no doubt studied the red ensign. When he spoke again, his voice carried no surprise. 'Come aboard, let us talk.'

'I will come aboard, sir, but it will not be to talk. We mean to have you in custody, whether living or dead is of no matter to us.'

They were closer now, and could see Thatch more clearly. He had a bottle of spirits in his hand and he swigged from it, before bellowing, 'Then you must come ahead, for damnation seize my soul if I will give you quarter, nor expect none from you.'

Thatch's black flag was then raised and the *Adventure* began to float into the swirling mist on a strong current, without taking the time to weigh their anchor.

'They've cut their cable,' Maynard said, looking up at the sails. 'Catch what wind we have, Mr Willard, while we have it, for we'll not let him slip away.'

'Mr Baker intends to attempt to block his course, sir,' said the first mate, nodding towards the *Ranger* which had engineered itself into the vanguard position ahead of them.

Maynard saw the manoeuvre and yelled across the water. 'For God's sake, Baker, keep out of reach of their guns!'

The mist swallowed both vessels, Maynard now anxiously staring at his sails, which only barely billowed. Flynt felt the tension rising among the men around him as they strained to see through the mist, each man hoping to catch sight of either ship floating out there in that deep stillness. But they saw nothing through the drifting, swirling vapour, heard nothing save the thud of rope on wood, the clink of metal, the creak of the *Jane*'s hull as it made its slow progress, the lap of the water below. Hands clutched weapons tightly, gunners readied themselves for action, seemingly frozen over their cannon as they watched and waited.

And then came the eruption of guns from deep within the murk, the muzzle flares briefly illuminating the mist, then dying. Timbers crashed, men screamed, the guns roared again. Men cursed softly, concerned for shipmates aboard the *Ranger*. Maynard's jaw tightened as he swung his telescope to and fro, jumping from one flash to another in a desperate attempt to see something, anything, that might illuminate how the *Ranger* fared in the engagement.

More cannon fire, more screams.

Then all was still.

'Who won, do you think?' Cain breathed, eyes narrowed as he searched the white curtain hanging before them.

'Thatch,' Flynt said.

'How do you know?'

'I know.' He sensed it. Thatch was still out there.

'Did the *Ranger* sink?'

'He's in the business of taking ships, not sinking them. He'll have lacerated it, killed as many as he could, then he'll leave it, retrieve it later as a prize.'

'And come for us?'

'And come for us.'

As if in sympathy with its companion vessel, the sails of the *Jane* at that moment stilled and flattened.

'We've lost the wind,' Maynard shouted. 'Break out the oars!'

With no time to mourn any lost comrades, the men rushed to heft seven pairs of oars, thrust them through the gunports and then bent their backs to give the *Jane* what speed they could. The only advantage was that if they had lost the wind, so had the *Adventure*. Flynt pounded up the ladder to the quarterdeck, Cain at his heels.

'He wants you to move closer,' Flynt said.

'Then I shall not disappoint him,' Maynard said.

'This is folly, Lieutenant. You've lost whatever surprise you had hoped, you've lost the *Ranger*...'

'We can't be certain of that.'

'I am,' Flynt insisted. 'Mr Baker was no match for Thatch, you must know that.'

'Mr Baker was...' He stopped and corrected himself. 'Is an officer in the King's navy and as such is more than a match for...'

'Goddamn it, Maynard,' Flynt growled, 'Thatch is a seasoned sailor. He learned his craft *in* the bloody navy. Pull away now, before it's too late.'

'Mr Flynt, I have my orders and my duty...'

'Do your orders include sacrificing your men and your ship?'

Maynard pulled himself fully erect and his jaw tightened. 'Damn you, sir, do not presume to question me, not on board my own ship. You may be a civilian but I can still have you in irons below. If your courage deserts you, then you and your companion are free to go over the side and swim to the island. But we will pursue this man and we will have him, on that you can be damn sure.'

A warning cry from the upper deck came almost too late. The *Adventure* swam out of the mist, so close it was as if they could reach out and touch it, and Maynard, Flynt and Cain barely managed to throw themselves onto the deck before its guns crackled a broadside, the small-calibre cannon blasting the deck with grapeshot, muskets peppering the hull. Mr Willard was nearly sliced in half, fragments of wood were ripped away to become projectiles themselves, one piercing the chest of the helmsman. On the deck below, men reeled as musket balls, shot or shrapnel found them. Legs were ripped out from under, arms pulled free from shoulders. Everywhere there was blood and screaming and smoke drifting from gun barrels and the acrid smell of burned powder.

Despite the withering fire, the surviving gun crews fell to their duty, the sharp report of the swivels as they raked the deck of the *Adventure* joining the din, shattering the rail of the ship opposite before it was gone, back into the mist. As the memory of the roaring ordnance died, screams and groans took its place and Flynt raised his head carefully. A gash on Maynard's forehead bled profusely and his eyes seemed to sail a little.

Cain was unharmed and he half rose to peer over the rail towards Thatch's ship. 'He's playing with us,' he said, bitterly. 'Did you ever feel so helpless, Jonas?'

Flynt didn't want to dwell on that. 'What are your orders, Lieutenant?'

Maynard seemed a little stunned as he dabbed at his bloody temple, wincing as he touched the wound. 'My orders?' He looked at the dead around him, at the blood and body parts, at men dangling from the rigging. The sailors left standing waited for him to tell them what to do. He shook his head to clear it.

'Maynard,' Flynt said sharply, 'what do you think he will do next?'

'Next?' Maynard swallowed hard, closed his eyes briefly, and when they opened again, some sharpness had returned as he assessed the damage. His eyes flicked over the dead and the dying again, his gaze not cold but professional, this not being the time for compassion or grief. 'He will believe he has decimated us and will come alongside, board us. We must maintain that fiction.'

There was little more than a dozen left standing out of the thirty-five crew, so that wasn't far from the truth, but Flynt guessed what Maynard's plan was.

'If you would assist me in getting the men below, I would be grateful,' Maynard said, staying low as he made for the ladder. They passed the word in whispers, telling them to keep below the level of the gunwale and to ensure they were armed.

The deck cleared quickly, each of the men vanishing through hatches or companionways. Maynard still seemed a little groggy but he remained on deck until the last, ordering Flynt and Cain to precede him below. As Flynt lowered himself through a hatch, he glanced back to see him kneeling on the deck, keeping watch for the return of the *Adventure*, but the mist was impenetrable. He crawled along the upper deck to the nearest companionway and joined his men, where in the darkness he whispered his orders.

'I do believe they mean to come aboard, and we shall allow them to do so. When I give the order, I want you all to swarm topsides once again, but by God, I want you screaming like demons. Give them hell, lads, for our shipmates on the *Ranger*, for our shipmates lying dead above, and for ourselves, for if we are not victorious this day then we will all be joining them.'

They listened to the sounds around and above. The ship lurched as something bumped into the keel, presumably Thatch coming alongside, then the sound of items rolling on the deck above and a series of small explosions.

'Grenades,' Maynard whispered. 'Filled with small shot, pieces of iron. The powder will burn but shouldn't cause any further damage to the deck.' He glanced around, cocked his head. 'The keel is sound, we haven't been holed. The ship will sail.'

'He'll mean to take her as a prize,' Flynt said.

'He will be disappointed,' Maynard said.

The thud and thump of the grenades ceased and the silence that followed was in many ways more disturbing than the roar of the guns. Now and again someone groaned above and each time, in the dim light cast through a tear in the hull, Flynt studied Maynard's expression. He remained stoic but there was anguish in the knowledge that his men were suffering and he could do nothing for them. Cain leaned against a bulkhead, calm now that action was imminent. Flynt also appeared relaxed, but his stomach roiled like a heavy sea, and he surreptitiously wiped his moist palms on his coat.

Heavy feet landed on the deck above, then Thatch's voice. 'The crew in the main be knocked down, come aboard, let's root out the remainder and learn them the error of their ways. Find 'em, lads, and cut 'em to pieces!'

There were laughs and then further footfalls, but it was impossible to gauge how many men there were. Maynard held up a hand to steady his force, his eyes scanning the boards above as if he could see through them. Footsteps quickened on the deck and shots rang out as the wounded were murdered. Maynard's jaws tightened. Thatch had said no quarter, and he meant it.

'Now, men,' Maynard said and was first to push open the companionway hatch to launch himself onto the deck, cutlass in one hand, pistol in the other, Flynt and Cain close behind. The lieutenant put a ball into the nearest pirate, and slashed a bloody gash in the chest of another. Arrogant he was, but the man had courage. His crew surged from where they had hidden, yelling and screaming, cutlasses clanging, pistols firing.

It was impossible to count how many pirates there were, for they were made vaporous by the veil of mist mixed with the smoke that drifted from the blackened marks left by the grenades, but they were fewer than Flynt expected. The two forces came together in a confused and confusing melee of clanging metal and cracking pistols. The deck, slippery thanks to the moisture in the air and the blood of the fallen men, was treacherous and each fighter had to struggle to remain upright. Men groaned, fell, blood flew, spurted from wounds, dark against the white air. Both sides battled bravely and well, the knowledge that it was kill or be killed proving a great motivator.

Then Flynt saw Thatch. He stepped from the murk, a pistol in each hand, laughing as if he was at a Drury Lane comedy. That was not the strangest, most disturbing thing about him. Woven into the ends of his beard were tapers, which he had lit in order to give off thin tendrils of smoke that waved and danced in his wake before evaporating. Maynard had ordered his men to give them hell, but Thatch looked as if he had just stepped from that place.

Flynt made for him just as a pirate stepped in his way, his sword swinging. He ducked from its path, pointed Tact upwards and emptied it into his chest. He then discharged Diplomacy at another who pressed an attack. Both pistols spent, he drew his sword and looked for Thatch again, didn't see him, but then found him once more. He took a few deep breaths and clenched his jaw in a bid to keep his rage under control. Rage unchecked was a liability. Rage focused was an asset. He was about to close on him when he caught sight of Cain warding off one pirate while another moved in. Flynt sprinted across the deck, kicked the man's leg from under him and then slashed at him as he went down. Cain despatched the other, saw Flynt standing over the dead man and gave him a little wave.

Flynt sought out Thatch again, finally seeing him facing Maynard, each firing pistols. Thatch's struck the cartridge box on the lieutenant's belt, but Maynard's hit home, puffing red in the pirate's chest, but he kept his feet and, with a roar, launched himself, cutlass raised. Maynard trapped the lunge and then they were hacking and slashing at one another, Thatch still laughing. Maynard was forced back but he continued to parry and strike as best he could. Flynt tried to reach the two men, but each time he made progress he was intercepted by one of Thatch's crew and by the time he was clear, Maynard was down on one knee, visibly tiring, while Thatch still roared and struck a vicious downward blow. Maynard blocked it but his blade shattered at the hilt under the force. Thatch laughed and prepared to strike again, but Flynt threw himself across the deck to shoulder him to the side, slamming him into the quarterdeck bulkhead.

Thatch was at first surprised to see him, but then smiled. 'Flynt, you think you can best me?'

Flynt stooped to pick up a discarded cutlass, knowing his sword wasn't up to the task ahead. 'Time will tell.' He tested the weight of the sword in his hand. 'But look where you are. On board a ship that is not your own and facing a man you have wronged.'

Realisation struck Thatch then and for the first time since Flynt had known him he showed something that might have been fear, but it was only momentary, for he immediately rushed forward with a howl. Flynt didn't try to stop the lunge, but stepped aside, whirled and rammed the hilt of the cutlass into the bullet wound on Thatch's chest. The pirate captain bellowed as he pulled himself away. As they circled each other, Flynt noted other wounds streaming red on the man's body and, despite his hatred, felt a measure of admiration at how he remained on his feet.

'You can't best me.' Thatch's voice displayed no sign of weakness. 'Not with a cutlass. You were equal to Hesikia, but not to me.'

Flynt kept his distance. 'I don't need to best you. I just need to keep my distance. Those wounds will do for you soon, I think.'

Thatch grinned and glanced at his blood-soaked body. 'These are but scratches. They are nothing compared to what I will do to you, Jonas Flynt...'

He leaped again, and Flynt parried the lunge, swung one of his own, but Thatch dodged aside, thrust his blade and Flynt only just managed to deflect it from piercing his gut. He aimed a kick at Thatch's knee, which had it connected may well have brought him down, but he missed. Thatch laughed, raised his sword over his left shoulder and arced it towards Flynt's neck. He ducked, the blade scything the air less than an inch from his head, and jerked his sword upwards, feeling resistance as he buried it in Thatch's belly. Any other man would have succumbed, but not he, for he pulled himself free and struck Flynt's cutlass away, sending it flying, then with another almighty shriek he closed in again. Flynt caught the wrist of his sword arm, while with his other hand he sought the bullet wound again, digging his thumb deep into the crevice, causing blood to bubble and cascade over his hand. Thatch's mouth tightened, his yellow teeth gritted, but he made no sound as he pushed harder, his free hand gripping Flynt by the throat. Flynt dug deeper, tugging at the fabric of his coat and the flesh beneath, ripping the wound wider, and this time Thatch groaned. But still he held on, thrusting Flynt against the bulkhead, increasing the pressure on his throat while also bearing down with his sword arm. Flynt rammed his knee into his groin but even that had little effect. His vision clouded as the constriction of his throat began to make itself felt and he summoned up every ounce of strength, every bit of rage and grief as he drew his thumb from the bullet wound, balled his fist and pounded it into Thatch's throat. The man's thick beard acted as a cushion, so Flynt's blows were almost useless. Flynt felt helpless as the pressure grew and all he could do was slap the heel of his hand against the bullet wound, but he was weakening. His breath was blocked, his muscles slackening, his eyes closing...

And then Thatch stiffened and howled and Flynt was free to collapse to the deck, coughing. He opened his eyes to see him backing away, one arm reaching behind him over his shoulder, the other stretching round his waist. He spun, his howl a scream, as he attempted to pluck out a cutlass buried deep into his back. Maynard was on his knees but falling back as Thatch flailed, his own blade swinging blindly.

Flynt leaped to his feet, snatched up his fallen cutlass, raised it high and swung it hard, putting into the stroke every muscle and sinew in his arm. The sharp edge caught Thatch across the neck, carving a deep gash and sending a fountain of blood erupting onto the deck. He choked, his sword slipping from his fingers. His hands clasped his throat as if he could stop the bleeding. He staggered backwards a few paces. Stopped. Remained still for a moment, giving Flynt a look filled with wonder at how it had come to this or whatever he believed lay ahead.

And then he pitched forward.

Around them the fighting ended. The remaining pirates threw down their weapons and the naval crew, what was left of it, herded them into a tight little group. Maynard pulled himself to his feet and stood over Thatch's lifeless body, his hand clutching a wound to his shoulder.

'Thank you, Lieutenant,' Flynt said, his breath sharp.

Maynard acknowledged him with a quick nod. 'Throw this filth over the side.'

Flynt moved forward. 'Not all of it...'

Epilogue

The myth of Blackbeard had begun while he was alive. On his death it grew. In the time it took Flynt and Cain to return to New Providence, stories of the pirate's death had reached ridiculous levels, with claims that when his body was thrown overboard it swam around the ship before disappearing. Given that Flynt carried his head in a wooden case lined with thick wool, that seemed unlikely. Directly on coming ashore, he took it to the fort, while Cain headed into the township to reunite with Anne Bonny. He had gone some time without a tupping, although he doubted she had. Flynt knew that such infidelity would matter nothing to him.

Woodes Rogers had recovered from his malady, but others on the island were not so lucky, and the small graveyard outside Nassau was stippled with fresh wooden markers. He was seated at his desk and peered only once into the box, before closing it again.

'I didn't think you to take me literally, Serjeant,' he said.

Flynt said nothing. Even though Thatch owed him a life, he was appalled by his thirst for blood.

Rogers waved for his secretary to carry the box away, which he did, his face wrinkled with disgust and holding it at arm's length, perhaps fearful that blood might seep through and stain his hose. Thatch had long since bled his last, though.

'What is for you now that you have had your revenge?' Rogers asked. 'You will return to London?'

Flynt had considered his future on the voyage back from Charleston, where they had found a ship. 'No.'

'Then you shall remain here?' Rogers was pleased. 'That is capital! Thatch is not the end of the pirate scourge – there are others still at large, Charles Vane being one...'

Flynt cut him off. 'I'm no pirate hunter.'

'No, but you are a hunter.' He indicated the door through which his secretary had taken the box. 'Had I not known it from Nat Charters already, that would have told me so. A pirate is just a cutthroat on a ship, no different from the rogues you deal with in London.'

'I'm one of those rogues.'

Rogers accepted that with a tilt of the head. 'Quite so, but you are a rogue on the side of the angels, are you not? And Toby Hawke and his son remain at large. I believe they still have a butcher's bill outstanding with you.'

The body of Hawke the elder, wherever it was, had still not been found then. As for his son, he had not been among Thatch's men captured on the *Jane* and awaiting trial and execution, nor was he among the few dead on the *Adventure*. That was an account Flynt fully intended to settle.

'I don't feel as though I'm on the side of the angels,' he said. 'Just my own, and I want a new life.'

'And you can have one. This is a new world, Serjeant, brimming with opportunity. Work with me, let us make this island civilised together. I need a man such as you at my back. Your friend, too, that fellow Cain. I understand he is a most useful man to have at your side.'

'I regret that I must decline. There is nothing for me in London, there is nothing for me here. I will see something of this new world, I think. As for Gabriel, he will return to England.'

Rogers accepted this with a sigh. 'Ah, well. There is work for you here if you change your mind. I thank you for what you have done. You have done your king a great service.'

Flynt placed his hat on his head, adjusted it. 'I didn't do it for him…'

—

Two weeks later Flynt and Gabriel Cain stood together on the dock and made their goodbyes. Cain didn't try to convince Flynt to return with him and Flynt didn't try to make him remain.

'Have you decided what you will do on your return?' Flynt asked as they watched a sack containing Cain's spare clothing being lowered to the boat that would row him out to the waiting ship.

'I will never be short of amusement or diversion, Jonas,' Cain said.

'And what of the Wraith?'

Cain's smile was half hidden. 'I told you before that the creature didn't exist.'

'But if he did?'

'If he did, he is dead and gone.'

'Never to return?'

'We should never say never, my friend. What of you? Will you ever return to merry old England?'

'I don't know,' Flynt said, and there was truth in that. He didn't know what lay ahead for him.

'You have people waiting for you. The beautiful Belle St Clair for one.'

Regret jolted him. 'I think not, Gabriel.'

'Cassie then?'

Flynt stared towards Hog Island. 'I bring her nothing but misery, Gabriel.' He faced his friend again. 'Will you go to Edinburgh?'

The oarsman asked Cain to hurry along. 'I very likely shall, at some point,' Cain said.

'Take care of her, if you can.'

'If she will allow me. As a friend only.'

For that, Flynt was grateful. He held out his hand. 'Safe travels, Gabriel.'

Cain took it and then clasped his other over the top. 'This is not the end for us, Jonas. It will always be thus for us, our paths diverging, intersecting, then diverging again.'

They faced each other, neither speaking. They had been through a great deal together. Flynt had never felt as close to any man as he did to Cain. He had saved his life more than once and had always been stalwart, if sometimes a little untruthful, especially over his past life as the Wraith.

Suspicion narrowed Cain's eyes. 'You're not going to embrace me, are you?'

Flynt laughed. It was the first time he had truly laughed in weeks. 'I wasn't thinking on it.'

'I'm relieved,' Cain said, 'for I save such attention for the ladies.'

They unclasped hands and Cain picked up his traps. 'Until next time, my friend.' He lowered a leg over the ladder leading to the waiting boat. 'Anyway, you have a debt to repay. You still have to save my hide.'

Flynt smiled. 'Did I not do that on board the *Jane*?'

Cain dismissed that with a hiss as he began to descend. 'That was only one instance, and I believe I could have taken that fellow. When I have leisure on the voyage home I shall calculate the full tally. I assure you it will be a long one.'

Flynt watched the boat pull away into the channel, Cain standing in the stern. 'Try to survive without me, Jonas,' he shouted.

'I will endeavour to do so, Gabriel.'

He remained on the dock until the ship sailed. As it headed out of the channel to sea he couldn't tell if Cain remained on deck.

He didn't need his glass to see; he knew he was there. Flynt knew that there were two things that would always be there. One was Gabriel Cain.

The other was regret.

Historical Note

This story began with a casual thought while having lunch with my then editor Kit Nevile. We were discussing the future of the Jonas Flynt/Company of Rogues series – the first one hadn't even hit the shelves – and I said that perhaps one day I would have Jonas heading to the Caribbean to not only encounter Blackbeard but also to kill him. Kit really liked that idea, but we had to wait before chronologically it made sense. And now here it is.

According to some sources, Blackbeard was actually killed by a Scotsman. He fought Lieutenant Maynard on the deck of the *Jane*, and may have won, had not a Highlander swung his sword and delivered a fatal neck wound. It was a short leap of my imagination to put Flynt on board, although Blackbeard's head was not transported back to New Providence, but was hung from the bowsprit of Maynard's ship and taken to Virginia.

The action near Ocracoke Island happened pretty much as I describe it, though I have embellished it for dramatic effect. This is a work of fiction, not a textbook. There was no mist, for instance, though the navy ships did run aground on a sandbar and had to ditch ballast to float free, and Maynard did order his men to row the sloop closer to Blackbeard's vessel. Even part of the dialogue between Maynard and Blackbeard is based on the reported exchange between them.

Blackbeard, of course, lived and breathed. His true name may have been Thatch, or Tatch, or Teach, or even Drummond. He may have had a family at home in England. He may have married bigamously many other women. He may have had syphilis. Whatever the truth, he was a larger than life character and I hope my version does him justice.

Woodes Rogers, Benjamin Hornigold (the architect of the Flying Gang), Charles Vane and Edward England all existed, although as usual their portrayal herein is my vision of them.

There was a pardon offered, with a deadline. Hornigold did accept and later became a pirate hunter. Vane didn't and escaped Nassau by setting fire to one of his prize ships and directing it towards the navy ships at the mouth of the channel.

Anne Bonny, who in my story has a sexual entanglement with Gabriel Cain, eventually became a pirate, too, after taking up with Calico Jack Rackham, but that was after the events I've used in my narrative.

Blackbeard had a second-in-command called Israel Hands, which I've amended to Hesikia (one of the names he was known by). I did this because I wanted to use the *Treasure Island* connection, so I made Israel younger and a friend of John 'Barbecue' Silver, who in the story did sail with Captain England as a young man. Robert Louis Stevenson's book is one of my all-time favourites; it used to be read to me at bedtime, and I couldn't pass up the chance of making reference to it. Especially given the name of my protagonist, even though it is a different spelling to the novel's notorious Captain Flint.

Captain Johnson also existed, but probably not as I paint him. A Captain Charles Johnson wrote a book called *A General History of the Robberies and Murders of the Most Notorious Pirates* (catchy title). It was an instant bestseller, recounting the lives and deaths of men such as Blackbeard, Hornigold, and others. For many years it was thought that the true author of the much-reprinted tome was none other than Daniel Defoe, but more recently that has been discounted. Nobody is certain who Captain Johnson was, or knows his background, and that mystery plays directly into my hands to make him a seasoned mariner with a past he regrets and a dislike for pirates.

I have telescoped the timeline of events down a little. Blackbeard was tracked down and killed in November 1718. Woodes Rogers didn't arrive in New Providence until the summer. For reasons of pace, I've fudged such details. I'm writing an historical adventure thriller here, so I make no apology.

The Nassau dignitaries, Taylor and Walker, were also based on real individuals on the island at the time.

The pirate stronghold in Nassau was deemed a republic and did have its rules, just as pirate vessels had their articles.

Acknowledgements

Thanks are due to the many people who have helped me with this series.

Author and friend the late Denzil Meyrick was the one who convinced me to embark on historical fiction in the first place, aided by our mutual agent, Jo Bell. Thanks to them both for talking me into it.

Kit Nevile has now gone on to pastures new, but while at Canelo it was he who saw the promise of Jonas Flynt and the Company of Rogues and helped guide it into reality.

Thanks also to Louise Cullen, Craig Lye and Alicia Pountney at Canelo for their sterling work for the book and series, as well as editor Hannah Boursnell and copy editor Jenny Page, and to Kate Shepherd for her labours on the publicity front, which is never easy in such a crowded market.

A big thank you to Isla Coole, of Criminally Good Books in York, for her incredibly generous winning bid in the Children in Read auction 2023. Along with a signed book, she also won the chance to have her name, or the name of a loved one, used as a character in this book. She gave that opportunity to her mother, whose maiden name is used. I'm sure my Madeleine McRobert is nothing like the real one, and I'm sorry she meets such a sticky end.

As usual, my author friends deserve a mention – Caro Ramsay, Michael J. Malone, Gordon Brown (not that one) and Theresa Talbot. They're always there for advice and kicks up the backside – whatever is needed. Usually the latter. Denzil Meyrick, unfortunately, is no longer with us but I'm sure he's watching from somewhere, always ready with a caustic comment.

Huge thanks to all the booksellers who not only stock the books but allow me to come in and scribble across them (yes, that includes you, Caron MacPherson of Waterstones Sauchiehall Street!). And let's not forget the reviewers, bloggers, festival show runners who have supported me over the years, not to mention the readers. You've all been great and everything you have done is hugely appreciated.

Finally, to my wife Sarah, who reads these things first and is my staunchest supporter and, when needed, deliverer of some gentle criticism. She also gives me my own shelf in her award-winning bookshop, The Book Nook, Stewarton. No, that's not why I married her.

OK, I've only one final thing to say…

JONAS FLYNT WILL RETURN!